A Short History of Geomorphology

Keith J. Tinkler

CROOM HELM
London & Sydney

©1985 K.J. Tinkler
Croom Helm Ltd, Provident House, Burrell Row,
Beckenham, Kent BR3 1AT

Croom Helm Australia Pty Ltd, First Floor,
139 King Street, Sydney, NSW 2001, Australia

British Library Cataloguing in Publication Data

Tinkler, Keith
 A short history of geomorphology.
 1. Geomorphology –History
 I. Title
 551.4'01   GB400.7

 ISBN 0-7099-2441-0

Printed and bound in Great Britain by
Biddles Ltd, Guildford and King's Lynn

CONTENTS

Part I

Part II

CONTENTS

CONTENTS

CONTENTS

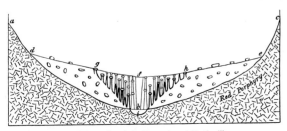

Diagram illustrative of the Formation of Earth-pillars.

LIST OF FIGURES, and other illustrative material

FIGURES

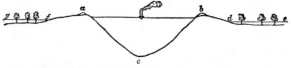

Section of Channel, Bank, Levees (*a* and *b*), and Swamps of Mississippi River.

Other illustrative material

Cover Design

Specimens, see Figure 7
Silhouette of an erratic, based on Figure 25 in Hull (1878)
Type is Old Face Baskerville, an eighteenth-century fount

To Shara

For her continual support and companionship

View near Llanwrtyd Wells, from a drawing by Mrs. Traherne.

PREFACE

I have written this book to fulfill a need which exists for a short yet reasonably comprehensive history of geomorphology. I do not neglect the monumental volumes being produced by Chorley, Beckinsale and Dunn (1964, 1973), but the first of these volumes is now out of print and the second is devoted, very necessarily, to a biography of W.M.Davis. The complementary third volume on Davis's contemporaries, and the fourth volume on the period since 1945 have not yet appeared (at the time of writing). The two existing books are excellent reference volumes, and all geomorphologists, including this one, are greatly in debt to them, but they are daunting and expensive to the hard-pressed student who needs a brief guide to the main ideas.

The other primary source for the history of geomorphology; G.L.Davies' 'The Earth in Decay,' is also out of print. Although this book confines itself to British geomorphology in the period 1578 to 1878, its scope is much broader than the title might imply since British geology held a central position in western thought until late in the nineteenth century.

There is one other monograph of which readers should be aware; Cunningham's 'The Revolution in Landscape Science' (1977a) which studies the critical transition years for geomorphology, from the late eighteenth to the early nineteenth century, with especial reference to the intellectual ferment of that period.

I do not intend to be pedantic in my use of the word 'geomorphology' which only came into existence at the end of the nineteenth century. (In Chapter 1 I cite a possible early use of the word by Naumann (1858) on the basis of a passing remark by Roglic (1972). However, I cannot find the word in the only edition of that date available to me through inter-library loan in North America.) It describes a field of interest that has always been clearly distinguished from other interests in geology and so I have used the word freely to avoid circumlocutions. Because geomorphology is indisputably a part of geology I have used 'geologist' as a generic covering term, even though I realize that because of institutional affiliations, geomorphologists are often geographers. Earth science has been used as an alternative to geology and is a useful term for the period when the subject of geology was not clearly separate from other surficial investigations. The meaning intended should be obvious from the context and there is no hidden significance in the use of particular terms, other than avoiding misrepresentation.

During the last two decades there has been a growing interest in the history of the geomorphology which parallels a similar interest in geology. The respect owing to the founders of our subject has grown to such an extent that Hutton and Playfair have been awarded posthumous knighthoods by the editors of one recent symposium volume (Coates and Vitek 1980)! The inauguration in North America during 1982 of a society and journal for the study of the history of the earth sciences will provide a welcome forum for discussion. Access to early literature has improved materially in the last twenty years. A great many early classics have been issued in facsimile reprints, and a number of translations have been made of key continental texts. Within particular branches of the subject the appearance of volumes of collected papers, linked by critical editorial commentaries, has made the task of analysis very much simpler.

As I shall demonstrate, the history of geomorphology has remained strongly rooted in national schools of thought and regrettably little transfer has taken place between them. For this reason this book inevitably has an Anglo-American bias, and a more complete history would read, in fact, as largely disjoint accounts of national schools.

Because of the constraints of length I have chosen to illustrate particular themes by studying certain papers and topics in some depth in an attempt to convey an overall view of the subject without sacrificing attention to detail. To readers who find some favourite paper passed over I should remark that in trying to see the subject as a whole, I have found that papers which are seminal to a particular branch of the discipline may have no impact on general thinking, or on the solution of a contemporary problem. This problem is particularly acute in the nineteenth century when many very 'modern' ideas were published, to be neglected for a century or more. I have tried to correct this bias on occasions, especially in the twentieth-century chapters, by citing somewhat anachronistically, these seminal but neglected nineteenth-century papers. Throughout the text the focus is upon the twin topics of process and form and their implications over cyclic time. In consequence much interesting material on chronology, stratigraphy and technique necessarily has been omitted, although it has not been neglected completely.

The balance of the book requires some justification although to a great extent the material organizes itself. I assume in the first instance that the reader of this book will have had at least one course in geomorphology to provide a passing acquaintance with terminology. With undergraduate students in mind, I have provided a brief introduction pointing out some of the problems inherent in writing and reading a history of geomorphology.

Chorley et al. (1964) deliberately dismiss the Classical writers since their subsequent influence on our subject was virtually nil. However, as I explain in Chapter Two, I feel that there are valuable lessons to be learnt from the Ancient World, not least because of its value as a comparison to western thought since the Renaissance. The second section of Chapter Two provides a compressed account of the linking period to around 1800, with the exception of the overlapping material in Chapter Three.

Davies (1969) has remarked that:

> In view of the importance generally ascribed to Hutton's theory, it is surprising that it should have received so little scholarly attention.

There is no space to rectify this deficit completely, which holds as much for Playfair as for Hutton, but I have devoted the bulk of Chapter Three to a review of the sequential development of Hutton's thinking, insofar as it is available to us. Playfair is justly famed for his 'Illustrations' of Hutton's theory, but as that volume is readily available in facsimile reprint, and is so succinct and modern as to defy ready summary, I have devoted the available space to the contemporary impact of Huttonian thinking, rather than to the book itself. Playfair, without doubt, deserves a critical modern edition and a biography in his own right.

I have made several references to the contents of John Playfair's library. It is a matter of great good fortune that a copy of the catalogue, compiled by the Edinburgh auctioneer John Ballantyre for the sale of Playfair's library beginning on January 20th 1820, has been acquired by Harvard University Library. The catalogue (Playfair 1820), which may be the only surviving copy, contains 1421 items and my comments in the text are based upon an examination of it.

The nineteenth century was the crucible from which modern geomorphology emerged and I make no apology for devoting over a third of the book to that period. Three chapters cover, in parallel, three main themes permeating nineteenth-century geology. The intermingling of these themes makes it essential for the reader to treat the three chapters as a coherent unit; catastrophism was not conquered, it metamorphosed into respectability. In addition, Chapter Four outlines the socio-economic backcloth to nineteenth-century

PREFACE

geological activity and the rise of North American geomorphology also is treated separately as it is otherwise difficult to disentangle from the systematic themes of uniformity, catastrophism and glaciation.

There can be no debate that Davis' Geographical Cycle deserves a chapter of its own although there is some overlap of themes with both of the subsequent chapters. Thus, in a similar vein to the nineteenth century, the three twentieth-century chapters should be read as group to achieve a balanced account. The division of the period since World War II into two, with a break circa 1960, is not intended to be dogmatic, and some themes have been allowed to run through to the present. Nevertheless, the date seems to be a reasonable one since the impact of the War and its aftermath had settled by then and subsequently the explosion of student numbers in the Universities (of which I was a part) catalyzed the new ways of thinking that were emanating from the United States.

The chapter on the period since 1960 proved the most difficult to write, mostly because of the enormous bulk of literature. It is impossible to give a truly balanced picture of the last two decades, especially as most of its literature is still active. For this reason a length commensurate with the quantity of literature is not justified: but it is not a critical problem. Whereas future historians of the subject will be able to see where current work is leading, we can only guess. All of my contemporaries will have their own perceptions about the meaning of what they are doing, and doubtless will guide their students accordingly. In so doing they will help shape the present, and through it the future.

The final chapter tries to illuminate the present with the wisdom and the waywardness of our predecessors, and to identify some persistent themes. It also attempts to anticipate the future by rooting in the past. Not to have made some statement on the future would have been an abrogation of the purpose of writing an account such as this, but whether the view expressed is right or wrong is, relatively speaking, immaterial.

Keith Tinkler.

St. Catharines

Department of Geography
Brock University

ACKNOWLEDGEMENTS

I am grateful to the following people for providing helpful information, both published and unpublished: W.E.H.Culling (London), Robert Dodgshon (Aberystwyth), Daryl Dagesse, Alun Hughes, Josephine Meeker, Peter Peach, David Rupp, Howard Williams (all of Brock University), Dr F.Ellenberger (Paris), the late J.Hoover Mackin, C.T.Stearns (Tufts University), Margaret Wilkes (National Library of Scotland) and Michael Woldenberg (Buffalo). Valerie Horvath helped collect modern bibliographic sources on the history of geomorphology and over several years the reference librarians at Brock University have located and obtained some very rare material and for their help I should like to thank Margaret Grove, Pat Paskey, Annie Relic and Barbara Whittard. Loris Gasparotto has been invaluable in advising me on methods of reproducing old illustrations to their best advantage, and for utilizing his cartographic skills. It is usual to thank one's typist but in this instance I can only pay tribute to the facilities of Applewriter I and II without which this book would have taken another year to produce, and to the Department of Geography and the Division of Social Sciences at Brock University for providing an Apple II+ and assorted printing facilities. However, as an unfortunate side effect all foreign material is bereft of accents, a deficiency that will irritate rather than hinder the reader. Dr Skilton of Brock University very kindly made available his program and expertise which greatly eased the preparation of the index.

Dr R.B.G.Williams (University of Sussex) has been for many years a stimulating source of conversations and ideas on the history of geology and has been generous in allowing me access to his library. Dr J. Menzies and Dr Lindsay Nakashima have willingly provided advice on glacial and coastal geomorphology respectively. My debt to the volumes of Chorley, Beckinsale and Dunn (1964, 1973), and of Davies (1969) will be clear to my readers and is gratefully acknowledged.

The following have graciously permitted me to reproduce

ACKNOWLEDGEMENTS

their material: The Royal Society of Edinburgh for Figure 7, also used in the cover design, The Geographical Journal for the headpiece to Chapter 9, Copyright © 1980 P.B.King and S.A.Schumm, and Geo Books for Figure 12, the Association of American Geographers Copyright © 1950 for Figure 13, Geological Society of America Copyright © 1932 for Figure 14 from Davis, W.M., 1932, Piedmont Benchlands and primarrumpfe, Copyright © 1960 The University of Chicago Press for Figure 15, 'The Fluvial System,' S.A.Scumm, Copyright © 1977 John Wiley & Sons Inc. for Figure 16, Professor M.J.Kirkby for Figure 17, the Honorary editor of the transactions, the Institute of British Geographers, Copyright © 1952, for the tailpiece to Chapter 10, the United States Geological Survey for the tailpiece to Chapter 9.

The President's Research Fund at Brock University has supplied funds that have enabled me to make extensive use of the excellent inter-library loan service available in Ontario, and North America. Harvard University Library kindly permitted me to examine the catalogue of John Playfair's library. I should also like to thank the following people at the University of Sussex for their hospitality during a sabbatical year spent at Sussex in 1980/81: The Dean of African and Asian Studies, Dr Chaudry, the Chairman of the Geography Subject Group, Professor T.H.Elkins, and the Director of the Physical Laboratory, Dr R.B.G.Williams. The University of Sussex Library kindly obtained materials on inter-library loan. The essential stimulus to write this book emerged in the relaxed atmosphere of Sussex during 1980/1, for which Brock University generously granted me a sabbatical year although not ostensibly for this purpose!

I should also like to thank the following people for their help in reading the text at various stages of completion: Daryl Dagesse, Elizabeth Williams, and Professors Lindsay Nakashima and Noel Robertson. An anonymous reader of the first draft made several suggestions that have materially improved the text and that help is greatly appreciated. My wife, Margaret, has read many versions of the text and in addition has had the laborious task of proof reading the final copy. If any errors remain after all this effort then I am tempted to remark, with Stephen Leacock, that they must be theirs!

Finally I must thank my family for allowing me to draw a substantial overdraft on the bank of family time in order to write this book.

Keith J. Tinkler

Part I

INTRODUCTION

To make a positive contribution one
could write a history of geomorphology.

(A. Hettner, 1928)

An episode in the history of the Glacier of Llanberis.

Chapter One

THE FRAMES OF REFERENCE

> Hamlet — *There are more things in heaven and earth, Horatio,*
> *than are dreamt of in your philosophy.*
>
> *(William Shakespeare, Hamlet)*

The process of understanding the geological writings of authors in previous centuries is affected by several premises which affect both the writer, and the reader. This chapter will attempt to illustrate them with specific examples, necessarily taken somewhat out of context. However, further details will be found in later chapters. In addition, I have tried to explain briefly what geomorphology is about, and how it is related to other subjects.

Under the heading of environmental constraints I try to emphasize an effect that is easily overlooked. Naturally, other sorts of constraint are also present; social, cultural, economic and technical, but I have chosen to discuss these as they arise in the subsequent narrative.

WHAT IS GEOMORPHOLOGY?

Geomorphology, as currently construed, is concerned with the shape of the earth's surface, more or less as we see it, and the processes that are involved in changing this shape over time. The subject falls most naturally within the province of geology although in much of the English speaking world, and especially the Commonwealth and former British Empire countries, geomorphology is institutionally associated with Geography rather than Geology as it is in the United States.

The global aspects of earth (geo..) shape (..morph..) science (..logos) implied in the word itself are not usually included in its study. The study of earth shape taken at the global scale is the province of geophysics in which subject it was already securely lodged by the time the word geomorphology came into general use after about 1890. There is, nevertheless, an interaction between the concerns of geophysics and geomorphology at the regional and the continental scale where land and sea beds are deformed by the varying loads placed upon them by water, ice and sediment and their corollaries, the melting of ice, the erosion of sediment from the continents, and the changing sea levels of the last two million years. It was not until well into the present century that the importance of these interactions was understood properly, and there remain many important problems to be solved in this field.

The word 'geomorphology' probably was used first by Laumann (1858) in the German language (Roglic 1972), but it first appeared in English in two papers by McGee (1888a,b) who attributed it first to Powell (1888b) but later claimed it for himself and credited Powell with the term 'geomorphic geology' (McGee 1893). General use of the word, in English, French and German probably followed its use by both McGee and Powell at the International Geological Congress of 1891. McGee spoke of the subject as novel, and as constituting the 'New Geology.' In addition it:

> is specially applicable in the investigation of the Cenozoic phenomena of the eastern United States and has been sucessfully employed in the region herein described; and, probably for the first time, important practical conclusions involving the consideration of hypogeal structure and orogenic movement, have been based on the interpretation of topography and on inferences from the present behaviour of the streams by which the topography has been determined (1888a).

But, if the word only appeared around 1890 did the subject itself exist in a recognizable fashion before this? The answer is definitely yes, particularly at the level of local detail. The earth's surface form is easily assimilated by the human eye, and the processes involved in effecting changes in this shape have been a focus of human enquiry and speculation since well before the days of Herodotus (5th Century BC). The visual impact of scenery is the most obvious facet of geology and must have been with man since the earliest days of his consciousness.

Even by the late eighteenth century, geology, a term introduced about 1687 by Warren (1690), and derived from the 14th century mediaeval Latin 'geologia', meaning 'the science of earthly things', was, in practical terms, either the description and meaning of scenery, with much debate on the impact of the Biblical Flood (i.e. geomorphology), or the experience of miners and quarrymen with minerals and rocks and which found an academic expression in mineralogy and palaeontology. During the nineteenth century, geology as a subject exploded in much the same way as, for example, biochemistry has exploded in this century. The explosion left geomorphology as a small part of a vast subject and with its emphasis on, or towards, the present it tended to lose touch with a parent subject so committed to exploring the past and unveiling the origin of the earth.

4

GEOMORPHOLOGY AND OTHER DISCIPLINES

Although geomorphology grew out of, and is a discipline within, geology, it has academic connections to several cognate disciplines. The processes of the atmosphere acting on the earth's surface over both the short term and the long term provide the essential catalysts that mediate the geomorphic system. On occasions geomorphology can repay the compliment by preserving, in transient morphology and surficial deposits, the records of past climatological events (Baker and Kochel 1982). With the essential energetics supplied by the atmosphere, it is obvious that a detailed knowledge of how these inputs vary in space and time is of paramount importance to geomorphology. Regrettably, meteorological records do not extend back more than a few decades in most countries and at most stations. In a few rare cases, records stretch back a few centuries, but not on standard modern instruments. The result is considerable uncertainty about the statistical record of climate, even over the Recent; the last ten thousand years.

The process of weathering at the land/air interface gives rise to a regolith from which the soil develops. The soil cover is, and has been, of the utmost importance to man as an agriculturalist over the last several thousand years. The soil, present or absent, is a fundamental control on how the morphology of landscape develops because the soil reveals an intimate and delicate balance between the forces leading it to accumulate and those of wash and wind that are tending to remove it.

A similarly intimate relation exists between geomorphology and vegetation, with soil and climate as important mediating agents. It is difficult to imagine Europe or North America as they looked a few millenia, or a few centuries ago and to realise the full extent of the forested cover that extended over both continents until forest gave way to tundra or to grassland. Nevertheless, the landforms we see have developed largely beneath a 'natural' vegetation that we no longer see, except perhaps in nature reserves. One great change in the character of landform evolution occurred when turf-forming grasses became a dominant vegetation in mid to high latitudes during the Tertiary. The resulting change in the nature of overland flow, which is severely inhibited by turf, probably converted many ecological deserts with sheetflow on extensive pediments into more undulating country with many small streams.

In our present day landscapes we find equally dramatic environmental metamorphisms, on the one hand in the deforestation associated with diffusing agricultural systems (Darby 1955), and on the other hand in the creation of that curious hydrological desert known to man as the City. Both have serious effects on fluvial processes which are not difficult to document and both serve to illustrate that the control on landscape processes exerted by vegetation cannot be under-rated.

At a more local scale, rates of re-colonization of newly-created shingle bars in coastal and fluvial systems are useful relative dating mechanisms (Hickin and Nanson 1975). Tree-ring chronologies, dendrochronology, especially using very long-lived species such as bristlecone pines are providing calibration of the Carbon-14 radio-active dating method that is unobtainable otherwise. In localized situations tree-ring counts can provide more precise dating than any other method, especially when applied to relative dating and short periods of time.

Very large scale geomorphology shares some common problems with geophysics. When sediment is eroded from a landscape over periods of millions of years, many thousands of cubic miles of sediment are removed from an area and transported horizontally to another area. The resulting changes of load on the earth's crust give rise to isostatic changes in level of the surface, measured with respect to a standard reference height. More localized loads, such as substantial lakes, volcanic cones, deltas or ice caps also give rise to deformations of the earth's crust. The importance of these effects to changes in the land level, measured with respect to a constant sea level, was not properly appreciated by geologists until the beginning of this century. The elucidation of the details of the regional deformations caused by, for example, the Pleistocene Ice Caps in North America and Scandinavia has depended substantially on the methods of regional geomorphology (de Geer 1892, Antevs 1922, Daly 1934, Andrews 1970). Geomorphological data helps shape geophysical theory and in return geophysical models can provide insights into isostatic changes of level for areas where no reference levels are possible, continental interiors such as Africa, and mountainous areas like the Alps, the Rockies and the Himalayas. In such areas isostatic changes are usually due to horizontal shifts of sediment, erosional loading and unloading, although it may be modulated by short term effects due to Pleistocene ice.

At the global scale, the almost universal recognition since the late 1960's that plate tectonics provides a global theory of geology means that, over time scales of the order of millions to hundreds of millions of years at continental to sub-continental spatial scales, geomorphology is having to take account of continental rifting, mountain building and sea-floor subduction within the scope of its own theories. The first geomorphological attempt to work at the global scale was L. C. King's 'Morphology of the Earth' (1962, 1967). King's book, which was firmly rooted in the work of Wegener, and the South African geologist, Du Toit, appeared just as 'Continental Drift' was being reborn in the guise of sea-floor spreading and plate tectonics. Not all geologists accept the validity of King's concept of world-wide contemporaneous erosion surfaces, but the notion that landform evolution since the Cretaceous needs to take account of plate tectonics is a fertile one and has already been applied to the Applachian region of the U.S.A. (Judson 1975). Doubtless the full impact of the need for geomorphological models sensitive to isostatic changes, particularly differential uplift and horizontal translation will not be felt until the next century: certainly current theories of landform evolution contain only a limited recognition of plate tectonics. Chorley (1963) provides, for the period before the adoption of plate tectonics, a fascinating review of the influence of changing geological notions of diastrophism on geomorphological thought since 1900.

Finally, hydrology and hydraulics are subjects with connections to geomorphology - especially that part of it dealing with the role of flowing water. Practical experience with these subjects goes back beyond Roman times since irrigation for agriculture, and the supply of water to cities, both require the construction of artificial waterways. The impact of these practical skills on geology and geomorphology was slight until well into this century although potentially it might have been much larger than it was. The construction of a 'regime' canal is a far cry from a natural river. The canal must transport water, but as far as possible not sediment and so, when constructed from local natural materials, a very delicate balance must be sought. In contrast, geologists have sought to understand how rivers are able to cut and maintain their channels, adjust to rapidly changing bed and bank materials, and still transport a varied load of sediment. Despite their opposing aims, both studies have much in common and it is all the more surprising that eighteenth and nineteenth-century engineers in France and Italy had worked out the basic hydraulics of canals and devised the notion of the profile of equilibrium at a time when many geologists denied that valleys were eroded by rivers, or even that they transported sediment to the sea (de Luc 1790, Kirwan 1799). By the time fluvial erosion of valleys was generally accepted, after about 1862, interest in geomorphology concerned landform evolution at the regional scale and over millions of years.

After World War II, a serious interest in the geomorphology of flooding in the United States led to a revitalization of the discipline by the infusion of ideas from civil engineering. Currently much work is being done at the interfaces of geomorphology, hydrology and hydraulic engineering and although the main emphasis in recent decades has been upon processes at short time scales of the order of one to one hundred years, longer timescales are coming into focus once more.

WHY STUDY THE HISTORY OF A DISCIPLINE?

The experience of conciousness is so rooted in the temporal domain that the loss of memory is counted one of man's most tragic afflictions. The loss of memory removes the basis of action and it must be a curious person indeed who sees no virtue at all in a modest acquaintance with what has gone before. There is a natural curiosity to search out the roots of modern thought, whatever the discipline, and it often comes as a shock to discover just how perceptive our predecessors were. Yet curiosity itself is only a partial justification. Far more important is the perspective gained on the environment in which ideas grow or wither, are rejected or discussed. As we move back before the present, we may be deceived, by the mere fact that we can read the language in which science was announced, into thinking that we understand the milieu in which it was embedded. Consequently we may acknowledge that a certain seventeenth-century writer described the process of surficial soil erosion, yet if that writer lived in an intellectual environment that placed no general significance on that fact, or even denied it, and if he also insisted that the Earth had only existed for six thousand years, we see that there is some dilemma in trying to assess his contemporary or his present importance to the subject (Harrington 1970, Gould 1970).

At any point in history it is important to try to establish the intellectual and technological constraints on contemporary thought, and from such a consideration we should be led to try to assess the nature and effects of modern constraints currently binding our subject. The difficulty, easily seen in hindsight, is that the constraints vary from age to age in their very character and are so much a part of the contemporary scene that they are not recognised for what they are. Prior to 1800 it is not difficult to see that the essential and rigorous constraint on geology was theological, bearing in particular on the age of the Earth, then generally assumed to be about six thousand years. In the nineteenth century the pendulum swung to the other extreme. Unlimited time and uniformitarian processes in the hands of Lyell, gave rise to the oxymoron that the essential constraint was the lack of a specific time constraint. There were also the compounding effects that (i) the landscape was usually construed to be the normal humid temperate landscape of high latitudes and (ii) the growing self-confidence and egotism of nineteenth-century physicists perhaps imbued others with the idea that problems that were 'merely' problems of applied physics were easily solved, at least in principle. The overall effect was that in the hurry to describe every local detail, the lack of a truly global system, as opposed to global principles into which the details could be put, meant that several large scale, observable, quasi-global effects were missed. In particular, isostatic compensation due to erosional unloading and sedimentary deposition, and the effects of ice loading and unloading during the Quaternary were almost completely neglected. Similarly the world-wide changes in sea level due to glaciation, first pointed out and calculated by McClaren in 1842, were mostly neglected until Daly (1910) laid the basis for understanding both isostatic and eustatic changes of level.

It is probably significant that it was in North America, where the scale is so grand and where subaerial processes are so active, that the first inklings of a comprehensive system of landscape evolution developed through the field work of Powell, Dutton, McGee and Gilbert and the masterly, if synthetic writings of Davis (1899). Yet, after almost a century of discussion on the Davisian model and its many derivatives, the debate about how landscapes evolve in time, or even if they do evolve in time, is still not settled. It is significant that Higgins (1975), in a review of this problem, called for a future geomorphic theory that is comprehensive and diverse, flexible and explanatory: in short, it must be able to accommodate under one head alternate modes of landform development.

Studying the history of the discipline will not in itself solve the problem or provide the theory but it may indicate in general terms the sources we need to draw upon and the scope of the scientific philosophy we have to encompass. It will also demonstrate the pitfalls that will likely be encountered in trying to propagate the new theory, and it will teach us that the essential stimuli may be voices in the wilderness, rather than prophets in our midst.

DEVELOPING A PERSPECTIVE ON WORKS OF THE PAST

It is not difficult to sit in judgement on the writings of the past: the difficulty lies in deciding on what basis the judgement should be made and to ensure, as far as is possible, that it is fair and reasonable. It is tempting to smile on reading that Sir James Hall (1761-1832), Hutton's close friend, thought that:

at some period very remote - with respect to our histories, though subsequent to the induration of the mineral kingdom, the surface of the globe has been swept by vast torrents, flowing with great rapidity, and so deep as to overtop mountains; that these same torrents, by removing and undermining the strata in some places, and by forming in others immense deposits, have produced the broken and motley structure, which the loose and external part of our globe everywhere exhibits' (Hall 1805, footnote p67-68).

9

It was to such events that Hall attributed erratics, unsorted superficial deposits and the excavation of valleys not caused by seismic fissures (Davies 1969).

Yet, insofar as anyone thought that the erosion of landscape was important, Hall was in line with the best contemporary views (other than those of Hutton and Playfair) and, it is well to remember, it would not be possible nowadays to defend the view that a typical Scottish landscape is entirely due to fluvial erosion given what we now know about Pleistocene glaciation during the last two million years. Of course, Hutton was right in asserting that the gradual processes of subaerial erosion which he observed could achieve significant work and that it could ultimately level mountains. He was largely mistaken if he thought that the erosion of the Scottish landscape, as we see it now, was entirely due to these processes.

How then shall we assess Hall's viewpoint? His attitude on the one hand appears to be a typical, 'conventional,' biblically inspired stand: on the other hand a careful reading of his later work (1815) shows that his intuition told him that something was wrong with Hutton's ideas at least as they applied to the Scotland he knew. For this he should be commended. Indeed, he went to considerable lengths to substantiate his deluge theory - by a literature review, field work, and by laboratory experiments (Hall 1815). In addition, he linked his theory quite specifically to the working of the Huttonian theory. In all these matters he was far ahead of his contemporaries. The example demonstrates that one should not adopt blanket viewpoints. Hutton's general principles were correct but his specific applications of them were sometimes in error, and his contemporaries should be given credit for recognising inconsistencies or ambiguities where they existed (Davies 1969, p196).

This establishes that there are at least two positions which must be considered: (i) The historical position is the assessment that we nowadays put upon a writer's work. To what degree did he correctly forsee our modern understanding of his problem? (ii) His contemporary position is established by asking how successful was he in explaining his position to his contemporaries and in converting them to his point of view. Both positions also require some degree of assessment of the rationality of any given theoretical viewpoint - to what degree is it based on contemporary science, to what extent is it a radical departure from the contemporary paradigm, and to what extent can the departure be justified?

The second item is very important in establishing the degree of influence that a writer had on the subsequent development of his discipline. For example, Hutton's principles of surficial erosion, and Playfair's illustrations of them, can be absorbed without much trouble into modern textbooks. However, Hutton's views on these matters were only partly accepted by his contemporaries, and he was only a little more successful in propagating his igneous theory of granite, as opposed to the Wernerian theory of precipitation. On the other hand, Lyell (1797-1875) was the arch propagandist for uniformitarianism: Hutton's gradualist approach to geological processes. He is remembered now for his overwhelming influence on nineteenth-century geology, and through it on our own century, whereas Hutton is usually given credit for first developing a comprehensive, and physical, approach to the subject.

A further difficulty arises in trying to understand what earlier authors meant, and in what sense they used some of their words. The danger lies in misinterpreting what was written and seeing the elements of a modern idea when it doesn't really exist. As an example, Davies (1969) refers to an 'enigmatic sentence' in Hutton concerning Alpine glaciers:

> The motion of things in those icy valleys is commonly exceedingly slow, the operation however of protruding bodies, as well as that of fracture and attrition, is extremely powerful (Hutton 1795, II, p296).

The difficulty here hinges on the meaning of the words 'operation' and 'protruding'. Hutton uses the word protrude elsewhere (II, p171) with reference to erratics:

> It is not always evident .. the particular route, which, in descending from a higher to a lower place, the protruding body had been made to take ...

Here 'protrude' means erratic, i.e. to be out of its proper use, to stick out, a meaning now restricted in geomorphology to rock fragments. The same sense can be applied to the glacier which 'protrudes' into lower valleys from its source in the Alpine accumulation zone. The British geologist Ramsay used 'protrude' in exactly this sense almost one hundred years later (1878, p370).

The word 'operate' is taken to have the meaning defined in the Shorter Oxford Dictionary as 'the working exertion of force, energy or influence.' The meaning of the paragraph can now be tackled. The 'icy valleys' are glaciers, which Hutton knew from the excellent descriptions in Saussure (1788), and which he discusses elsewhere, (Hutton 1795, II, p330), and so the sentence may be read to refer to the work done by the glaciers which extend from their high Alpine sources to lower and more temperate valleys below. However, it is most likely that in referring to the 'extremely powerful .. operation' of these glaciers he was thinking of the massive loads of surficial moraine that Saussure so clearly described, rather than glacial excavation, by subglacial mechanisms, of pre-existing fluvially eroded valleys.

11

This last point illustrates that too much should not be read into his meaning: statements must be read in the broad context of an author's writings and not with the intention of extracting a semantically feasible, but scientifically exotic meaning that excites the imagination but which is not based upon sober analysis.

ENVIRONMENTAL INFLUENCES ON THEORIES AND OBSERVATIONS

It is trivial to remark that what can be learned of a landscape is constrained by the rocks present and by the surficial processes that shape them into landforms. Nevertheless, it is certainly true that the British landscape, while ideal in some respects - teaching William Smith the principles of paleontological correlation and James Hutton the truth about granite and unconformities - is ambiguous on the issue of recent geological processes. Glacial drift, before the general acceptance of the glacial theory around 1865, was a very puzzling deposit. Clearly distinguished from the underlying rocks, it presented few systematic properties across wide areas. It was difficult to deny, in the absence of any knowledge of sedimentation, that it was truly the remnant of a great flood - usually and easily identified as the Mosaic Deluge by those anxious to placate theologians. Similarly, the relatively few signs of recent uplift in Great Britain (raised beaches) could not be correlated with recent volcanic activity or with significant historical earthquakes. The importance of constraints of this sort can be seen from the remark made by Lyell in the preface to Volume iii of his 'Principles of Geology' (1833):

> I occasionally amused myself with speculating on the different rate of progress which Geology might have made, had it been first cultivated with success in Catania where the phenomena above alluded to, the greater elevation of the modern tertiary beds in the Val di Noto, and the changes produced in the historical era by the Calabrian earthquakes, would have been familiarly known.

Lyell records how surprising this was to him and how it illuminated Descartes's precept 'that a philosopher should once in his life doubt every thing he had been taught.' However, he made good use of his experience by devoting, Chapter V of his Principles to 'Prejudices which have retarded the progress of Geology.' While some of Lyell's discussion is devoted to dispelling prejudices stemming from essentially theological philosophies and attitudes, one subheading discusses those 'arising from our not seeing subterranean changes.' It was an astute psychological step by

Lyell, who realized how much mental inertia there was to overcome in his more obdurate contemporaries, to reveal the philosophical basis upon which their objections were based before sweeping away their insecure foundations.

Ellenberger (1980) has examined the problem of environmental influences in the specific context of the French naturalists of the eighteenth century. He shows that those living in the southern parts of France realized more clearly than their northern colleagues the realities of tectonic deformation and the work of running water. In their turn, naturalists in northern France tended to interpret ubiquitous Pleistocene deposits as evidence of the Mosaic Deluge.

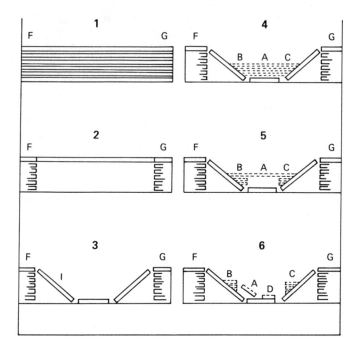

(1) Precipitation of sedimentary rocks into a universal ocean.

(2) Earth's pristine surface; development of subterranean caverns by sapping.

(3) Collapse of undermined continents, valleys drowned by the sea = Flood.

(4) New sedimentary rocks form in valleys, including fossils.

(5) New rocks emerge at the end of the Flood, undermining as in (2).

(6) More collapse to create the present topography.

Figure 1. *Steno's interpretation of the Tuscan landscapes. Redrawn and reorganised from Steno's* Prodromus *of 1669.*

Even further back in time the Dane, Nicolaus Steno (Figure 1, Steno 1669) appears to introduce fanciful notions into his series of diagrams explaining the evolution of the Plain of Tuscany. In stage 3 he calls upon the 'sudden up-thrusting of gases and the collapse of air-filled cavities' to disrupt horizontal strata into hills, mountains and valleys. It is well to remember, though, that much of Italy is a land of limestone with its associated caverns, of active volcanoes (Vesuvius, Etna), and of frequent earthquakes. Bearing in mind Lyell's remarks 150 years later, just quoted, it is more understandable that Steno, having enunciated the principle of horizontal strata being deposited beneath the sea should seek to convert it into hills, valleys and mountains. Seen in the environmental context of Italy, his scheme is much less ridiculous than it might appear at first sight. Indeed Davies (1969) has remarked on the pervasive influence that the writings of the Classical authors (based mainly in Italy and Greece) had on much early thought in British geology.

In the late nineteenth and early twentieth centuries similar constraints have operated. The 'normal' cycle of erosion developed by Davis (1899) was seen by the 1950s, and even by Davis himself in the 1920s, to need substantial revision when applied to the fluvial landscapes of semi-arid regions. Academic exploration of the tropics revealed a rich variety of landscape types: some involving the deep chemical weathering of rocks, the etchplains of Wayland (1933), others the planation of bedrock surfaces by brief torrential flashfloods to create the pediments and pediplains described by L.C.King (1953).

Nevertheless, in future centuries our successors will see that in the present century we have been limited to what the earth can teach us. Lately we have stood on the Moon and observed the martian surface closely, but satellite photography and grab samples are no adequate substitute for field work, and that will have to await the twenty-first century. The lunar and martian surfaces provide excellent environments with radically different temperature, moisture and gravitational regimes that have acted undisturbed for millions of years and without the interference of a recent Ice Age. It would seem logical that other planetary surfaces can teach us much, by inference, about the earth's own surface but that, in the truest sense, still remains to be seen.

More sobering still, if only we could conceive of them, would be the constraints about which we cannot even hypothesize: so strong is our myopia.

Part II

BIRTHPANGS OF A DISCIPLINE

Nunc naturalem caufam quærimus et affiduam, non raram et fortuitam.

SENECA.

Marble Column from Temple of Jupiter Serapis.

Chapter Two

THE ANCIENT WORLD, THE RENAISSANCE AND THE AFTERMATH TO 1800

> Str. – *But why do they look so fixed on the ground?*
> Dis. – *They are seeking what is below the ground.*
> Str. – *Ah! they're looking for onions.*
>
> *(Aristophanes, 423 B.C., The Clouds)*

When the finest library in the Ancient World, at Alexandria, was finally destroyed in AD 642 on the orders of Omar Caliph, (it was partly burned in 47 BC and again in 390 AD), the modern world was left with a literary heritage of fortuitously preserved manuscripts that fill but a few shelves in a modern library. Despite the almost random basis of our knowledge of ancient thought, it is still possible to get a fairly clear picture of the intellectual milieu in the millennium or so preceding the sack of Alexandria, and within which ideas on landscapes and landforms existed.

Because modern knowledge of landforms is not founded directly upon ideas descending from the Ancient World, even though there are notions in common, it may be argued that a study of their works is of largely antiquarian interest (Harrington 1967, 1969, 1970, Gould 1970). That attitude would, however, miss two points. Firstly, the Ancient World provides an entirely separate intellectual milieu with its own constraints on thought which conveniently illuminates how identical observational material may be treated very differently merely because of a different world view.

Secondly, and of equal importance, is the fact that the revival of learning in the late medieval period and during the Italian Renaissance was heavily dependent on the survival of Greek and Latin manuscripts, repeatedly transcribed, in the monasteries and libraries of Europe. These ancient sources quickly achieved an authority and secular sanctity second only to that of the Roman Catholic Church, but which would not have been tolerated in the Ancient World itself, where authors tended to be highly critical of their sources and of one another. In consequence, Classical literature, and the views of those authors whose writings happened to survive, exerted an undue and often unfortunate influence on the thought of many generations, not withstanding Roger Bacon's wish to 'burn all the books of Aristotle.'

Perhaps the most striking feature of the ancient literature is the general lack of progress or development,

17

from a theoretical point of view, in understanding landscape systems. Landforms were not, of course, segregated from other aspects of the natural landscape by the Greeks and Romans who had a holistic view of the world, and a lack of theoretical progress was general in natural science. However, this was not paralleled in mathematics, astronomy or civil engineering where modest advances did take place.

As a basis for discussion I will examine first the philosophical bases of Greek knowledge because the Romans adopted many facets of Greek life and attitudes once their domain extended over former Greek territories and there are no extant writings of interest after the first century AD. A philosophical approach will provide more understanding of the subject than a chronological one when the period is devoid of intellectual progress. Figure 2 provides a chronological view of the Ancient World with authors plotted according to their lifetimes and broadly classified by philosophical beliefs.

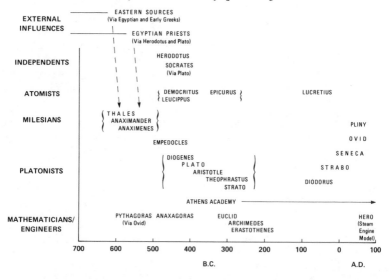

Figure 2. *Authors of the Ancient World who wrote about landforms, classified by lifetimes and philosophical beliefs.*

PHILOSOPHICAL BASIS OF NATURAL SCIENCE IN THE CLASSICAL WORLD

At the root of all Greek philosophy was a profound love of nature and their belief that the natural environment was both the playground and workplace of the gods, but respect for deities did not inhibit a natural curiosity about the physical workings of the world. Nor did they fall short in making suggestions on how the hidden mechanisms of the world might operate - e.g. what prevents the seas filling up or, how is river flow maintained in the absence of rainfall?

One can extract three major principles from Greek writings that would suffice as a basis for rational investigations of landforms. The first is the concept of infinite time - a universe eternal. The second is a belief in the reality of denudation - mass loss from the landscape. The third is an acceptance of the conservation of mass - nothing is lost and nothing is gained.

The tenet of infinite time is equivalent to modern day requirements for steady state theories of the universe, but the measure of its infinity is rather less important than the fact that the time available for the working of natural processes was not constrained in any obvious manner and in particular did not hinder speculation in the temporal domain, be it by hindsight or by prediction. Herodotus (c484-c426 BC), in his description of Egypt (The Histories, Book II-11), compares the alluviated lower valley of the Nile with smaller examples in Greece, the Maeander and Achelous valleys, and then turns to speculating:

> Now if the Nile choose to turn his waters into this Arabian Gulf, what hinders that it not be silted up by his stream in twenty thousand years? nay, I think that ten thousand would suffice for it. Is it then to be believed that in the ages before my birth a gulf even much greater than this [i.e. the Nile valley itself] could not be silted up by a river so great and so busy?

This position contrasts markedly with that in post-Renaissance Europe when the beginning of time was placed only 6000 years in the past and when the end of the universe was predicted within a few hundred years.

The reality of denudation is so self-evident, it would seem, that its recognition as a principle would be superfluous were it not for the fact that once more in post-Renaissance Europe there were those who sought to deny it. Descriptions of erosion are extant from Homer's 'Iliad' which probably dates back to the ninth century BC. The Greeks took natural catastrophes to be the wrath or revenge of the gods and so in the following quotation when Zeus had poured forth rain, then on the earth:

> rivers flow in flood, and many a hillside do the torrents furrow deeply, and down to the dark sea they rush headlong from the mountains with a mighty roar, and the tilled fields of men are wasted...

Lest we should doubt Homer's word, Strabo (64 BC - cAD 25) made a great point of defending the geographical knowledge of Homer, against Erastothenes (c276-c196 BC) in particular, calling him the first geographer 'at a time when the dawn of civilisation was opening the human mind' and inferring considerable changes in the Nile delta on the basis of incidental comments in Homer. Half a millennium later the remarks of Plato (427-347 BC) on soil erosion in Greece and Turkey can still be used as a documentary basis for modern stratigraphical studies of denudation in the first two millenia BC (Davidson 1980, Bintliff 1976).

Finally, the view that matter is conserved over the globe as a whole was to the Greeks, as it is to us, infinitely more preferable than a world in which matter could come and go at the whim of an author, at least until the need could be justified from the physics itself.

These three principles were known and taught by Pythagoras (c590-c510 BC) who, it was said by Plutarch, 'thought much, said little and wrote nothing.' Plato was probably influenced by the mystical and mathematical pythagorean school at Croton in southern Italy but his thought is best known through the brief and possibly anachronistic account given by Ovid (43 BC - 17 AD) in his 'Metamorphoses' (Book 15). Pythagoras is made to speak of the 'everlasting universe' and then to state that:

> Nothing is constant in the whole world. Everything is in a state of flux, and comes into being as a transient appearance. Time itself flows on with constant motion, just like a river...
>
> Nor does anything retain its own appearance permanently. Ever-inventive nature continually produces one shape from another. Nothing in the entire universe ever perishes, believe me, but things vary and adopt a new form.
>
> I have seen what once was solid earth now changed into sea, and lands created out of what was once ocean. Seashells lie far away from the ocean's waves, and ancient anchors have been found on mountain tops. What was at one time a level plain has become a valley; thanks to the waters flowing over it, mountains have been washed away by floods and levelled into plains.

The implicit conflict at any particular site on the earth's surface between the second and third principles discussed above was resolved for some thinkers by invoking a cyclic behaviour for the earth's surface. A notion of cyclicity goes back from Aristotle (384-322 BC) through Plato, and possibly Pythagoras, to the Egyptian Priests from whom he probably got the idea of the world renewing itself, or being purged, by floods or conflagrations. The Egyptians in turn may have got the idea from sources further east. Pythagoras hinted at this in the quotations given above, and in others not given here, but in the hands of Aristotle the notion is specifically explained with reference to actual examples. In his

'Meteorologica,' Aristotle disarmingly states that 'on this subject as on many others we know of no previous theory that could not have been thought of by the man in the street.' Despite Bacon's growl of disapproval quoted above it must be said that Aristotle is remarkably sound on the subject of earth processes, which is not say that he is always right. The following quotations (Book I-xiv) try to encapsulate the full range of Aristotle's theoretical thinking in order to do justice to a comprehensive scheme:

> The same parts of the earth are not always moist or dry, but change their character according to the appearance or failure of rivers. So also mainland and sea change places and one area does not remain earth, another sea, for all time, but sea replaces what was once dry land, and where there is now sea there is at another time land. This process must, however, be supposed to take place in an orderly cycle. The originating cause is that the interior parts of the earth, like the bodies of plants and animals, have their maturity and old age. Only whereas the parts of plants and animals are not affected separately but the whole creature must grow to maturity or decay at the same time, the parts of the earth are affected separately.
> But these changes escape our observation because the whole natural process of the earth's growth takes place by slow degrees and over periods of time which are vast compared to the length of our life, and whole peoples are destroyed and perish before they can record the process from beginning to end...For the land of the Egyptians, who are supposed to be the most ancient of the human race, appears to be all made ground, the work of the river.
> This has happened in Greece in the lands around Argos and Mycenae... What has happened in this small district may therefore be supposed to happen to large districts and whole countries.
> Those whose vision is limited think that the cause of these effects is a universal process of change, the whole universe being in process of growth. ...we must not suppose this is the growth of the universe, for it is absurd to argue that the whole is in process of change because of small changes of brief duration like these; for the mass and size of the earth are of course nothing compared to that of the universe.

Aristotle makes some very astute remarks in the course of these quotations although he is vague on the precise mechanism by which significant vertical changes of land and sea could take place, if indeed he thought they could, as opposed to the coastal erosion and river alluviation that he describes in his examples. He is clear, however, that there is a regularity and order to the process and envisaged that it was analogous to a season but 'at fixed intervals in some great period of time.' The returns of these great catastrophes were determined by the period of the Great Year, a theoretical cycle of the Solar System, and variously estimated at between 10,000 and 360,000 years long.

With such a set of principles, and the careful thought that lay behind them, the Ancients where in a far better position to begin a rational investigation into the surface workings of the globe than were most thinkers in Europe after 1500. Yet, despite this headstart, little real progress was made. Part of the reason for this probably lies in the systems of physics adopted by the Ancients, and the Greeks in particular.

THE PHYSICS OF THE ANCIENT GREEKS

It is probably incorrect to view the physics separately from the philosophy of Greece in Classical times, but it is convenient here to regard the physics as the basis upon which the mechanisms of natural processes would be founded: the basis of a mechanics of action. While Greek physics acted in this way for the Greeks it did so in a very subjective way and the scheme adopted was neither suitable for calculation nor for experimentation.

`The earliest Greek thinkers on record, the Ionian school founded by Thales of Miletus (c624-c565 BC) and his disciples Anaximander and Anaximenes, thought that the primary elements were water (Thales) and air (Anaximenes) in association with earth; ideas possibly influenced by the powerful operation of natural processes on a rocky coastline. These ideas developed into the notion that the four elements were earth and water, air and fire and as Ovid puts it in the words of Pythagoras (Metamorphoses, Book 15):

> Two of these, earth and water, are heavy and, by their own weight sink down, while the other two, air and fire, which is more rarified than air itself, are weightless, and soar upwards, unless something holds them under. Though these four elements are distinct from each other in space, yet they are all derived from one another, and are resolved back again into themselves.

This view dominated most thought in the Ancient World and particularly as it applied to geological processes. Of special importance was the idea of mutability - that one element could turn into another, usually as a result of the application of heat or cold. Seneca (4 BC - 65 AD), Stoic philosopher and Nero's tutor, attempting to explain river baseflow, argues that just 'as a change of atmosphere produces rain so a change of earth produces rivers' and specifically cites:

> the principle that all elements come from all others, air from water, water from air, fire from air, air from fire, so why not water from earth ... most especially water, for the two are related. Both are heavy, both are dense (Natural Questions, Book III-10).

According to Dampier (1966), the intellectual heirs of the Ionians were the Atomists founded by Democritus (c465-c375 BC) and Leucippus (c470-c400 BC), maintained by Epicurus (342-270 BC) and revived by the Roman writer Lucretius (95-55 BC). This school held that the world was nothing but fundamental particles, the atoms, and voids. While superficially closer to a modern view of the universe it was not founded on a system of mathematics, and could not be used as a basis for experimentation. It appears to have had relatively little influence on what may be construed as geological thinking, although it apparently led Democritus to believe that the sea was slowing drying up.

Early Greek philosophy was motivated by a type of animism that held that the earth as a whole was alive - a global organism - although the traditional polytheism often re-asserted itself. Eventually, however, the mechanical workings of the world became a subject of logical speculation largely devoid of deistic overtones, although the maturing process mentioned by Aristotle reveals traces of this earlier thought in Greece's greatest thinker.

None of the physical notions noted above were suitable as the foundation of true science although the popular scheme of four elements was ideally designed for subjective metaphysics. Regrettably the Greeks never really devised a scientific methodology close to that of our own times in which the pivotal step is the experimental testing of logical deductions based on observed data. Above all, the Greeks prized the idea of logical thought but despised those who would dirty their hands with field work. Only Socrates (469-399 BC), who left no writings but whose ideas are known through Plato, rejected all the rival theories of the philosophers and subjected every idea to rigorous testing. For his pains he was caricatured by Aristophanes in his comedy 'The Cloud,' but his strength was his failing, for his attitude was too eclectic to appeal to mainstream Greek thought - a flood of logical and critical thought maintained for centuries by Plato's academy in Athens: open from 385 BC (and open to women) until closed by the Emperor Justinian in 529 AD. It is reputed to have had as many as 2000 pupils at the peak of its popularity.

The lack of an experimental methodology is all the more

surprising in view of the fact that observational descriptions and inferences extant from Greek, and subsequently, Roman writers are remarkably accurate and reliable. For example Plato records in 'Critias' that:

> In consequence of the successive violent deluges .. there has been a constant movement of soil away from high elevations; and, owing to the shelving relief of the coast, this soil, instead of laying down alluvium as it does elsewhere, has been perpetually deposited in the deep sea around the periphery of the country or, in other words, lost ... All the rich, soft soil has molted away, leaving a country of skin and bones (so that rainfall) is allowed to flow over the denuded surface (directly) into the sea (Critias, 111, D-E).

Similar passages can be gleaned from many other Classical writers. In many instances, and especially the case of Egypt, documentary sources were quite sufficient to provide good temporal baselines as a basis for long term inferences. Records of the height of the Nile flood were kept as far back as the reign of Menes (3188-3141 BC) and the nilometer at Aswan, a stage recorder in modern terms (one of several along the river), was described by Strabo:

> The nilometer is a well on the bank of the Nile constructed of close-fitting stones, in which are marks showing the greatest, least and mean rises of the Nile; for the water in the well rises and lowers with the river. Accordingly, there are marks on the wall of the well, measures of the complete rises and of the others. So when watchers inspect these, they give out word to the rest of the people, so that they may know; for long beforehand they know from such signs and the days what the future rise will be, and reveal it beforehand. This is useful, not only to the farmers with regard to water distribution, embankments and canals, and other things of this kind, but also to the praefects, with regard to the revenues; for the greater rises indicate that the revenues also will be greater (Geography, Book 8-I-48). '

In spite of this rich observational basis at the regional scale there was little or no attempt to work at the micro-scale of process, even though much might have been done with the rudimentary technology available. Nor were observations made with the intention of testing ideas or refining previous findings. If Empedocles (c490-c430 BC), really led a hermit's life on the crater rim of Etna for several years and finally lost his life there, it is more probable that he sought mystical inspiration about the element fire than first-hand facts on vulcanology, although he doubtless learnt many. Yet Empedocles was no dreamer; he had at one time drained swamps near Selinus and blocked a windy rift in the hills to improve the micro-climate and harvests at Agrigentum in Sicily (Forbes 1957).

To be fair, the failure to develop a truly experimental science in Classical Greece probably had a technological basis. Lee (1952), in his introduction to Aristotle's 'Meteorologica,' notes a lack of precision instruments (accurate clocks), of good glass making, of sophisticated iron making - a prerequisite for good machines - and a mathematical system ill-adapted to scientific notation and cumbersome calculations. Coupled with disdain for manual labour, these lacunae effectively prevented any chance of true scientific method developing in the fields of physics and chemistry, from which they might have spread to the earth sciences. Significantly the Greeks progressed best with those subjects that minimise manual experimentation: medicine, mathematics and astronomy. Likewise, necessarily manual enterprises such as military and civil engineering maintained an existence independent of the philosophers, e.g. Archimedes (287-211 BC), although there was common ground in mathematics. Strabo, for example, mocked Erastothenes for relying on the measurements of engineers for his opinion that the Red Sea and the Mediterranean differed in level, rather than relying on reason. Erastothenes, he says in derision, 'has made engineering a branch of mathematics.'

An instructive example is Aristotle's analysis of stream flow (Meteorologica, Book I-xiii) in the light of the reservoir theory which held that stream flow was maintained from huge subterranean caverns filled with water:

> But it is evident that if anyone tries to compute the volume of water constantly flowing each day and then to visualize a reservoir for it, he will see that to contain the whole yearly flow of water it will have to be as large as the earth in size, or at any rate not much smaller.
> It is not, of course, at all impossible that there do exist such places containing large volumes of water, like lakes, but they cannot be so large as to act in the way this theory maintains.

Aristotle's conclusion is correct but it is perfectly evident that he never did the calculation, or if he did he was either sadly mis-informed about river flow or made serious errors of arithmetic. In any case it can be judged as an example of the failure of the experimental method. For the record, and using modern values for the flow of the Nile, and the Nile was a wonder of the Ancient World that attracted Greek intellectuals as the Grand Tour of Europe was later to attract the English aristocracy, it may be computed that the annual flow of the Nile could be contained in a cube of sides 4.48km. Twice this flow, i.e. allowing for measurement errors, would need 5.64km and the entire annual flow of the earth (most of it unknown to Aristotle) would require only a cube of sides 30.50km. Thus we have some measure of the magnitude of Aristotle's failure in this example.

A more trivial but nonetheless significant example is provided by one of the 20,000 facts that Pliny (25-79 AD) records in his 'Natural History.' He claimed 'that liquid is reduced in bulk by freezing, and when ice is thawed the bulk produced is not the same' - the reverse of the actual case and a fact not difficult to verify. Pliny, however, can be honoured for the manner of his death: he was overcome by fumes during a foolhardy field excursion to the shore during the 79 AD eruption of Vesuvius.

Pliny's encyclopedic but indiscriminate 'Natural History' forms an unsatisfactory punctuation mark to the record of earth science in the Ancient World, and it will be convenient to consider the more circumspect authors, Seneca and Strabo, in a parting perspective.

THE ROMAN CODA TO THE ANCIENT WORLD

The Romans were quintessentially a practical people, and while they acquired many of their aesthetic views from the Greeks, they developed engineering for their own purposes, and the Greeks had little to say on these matters. Alexandria still flourished as a centre for Greek science, but the simple steam engine developed by Hero (c70 AD) remained a temple toy and did not fuel the imagination of an ancient Hutton with thoughts of an earth machine. The lack of sympathy between Greek theory and Roman practice stunted both, and the modern world is left with the magnificently over-engineered public works of the latter and copies of long-perished manuscripts from the former. The Romans could supply Rome with 11.70 cubic metres of water per second, yet a leading engineer, Frontinus (c100 AD), failed to make volume measurements, being content to compare the cross-sections of aqueduct channels and to blame water losses on dishonest citizens.

While I have emphasized the lack of progress made in earth science over the better part of a millennium in the Ancient World, this is not to deny that the later writers owned a certain perspective due to their access to a varied collection of writers. Notable amongst these was Seneca whose work is contained in his book 'Natural Questions.'

For the most part Seneca is content to review the standard opinions within the physical framework of the four elements and considers the main problems of contemporary interest. One of these was the reason for baseflow in rivers.

Floods were happily attributed to rain storms, or melting snows, but it was a matter of continual interest:

> how the earth supplies the continuous flow of rivers, and where such great quantities of water come from. We are surprised that the earth is not affected by the addition of the rivers; it is equally surprising that the earth is not affected by the loss of waters leaving it (Book III-4).
> - as a diligent wine-gardener myself I assure you that no rainfall is so heavy it wets the ground to a depth of ten feet. All is absorbed in the outer surface and does not yet get down to lower levels (Book III-7).
> - the ground is either dry and uses up what is poured into it or it is saturated and will pour off any excess that has fallen into it. For this reason rivers do not rise with the first rainfall because the thirsty ground absorbs all the water (Book III-7).

I have already quoted Seneca on the mutability of the Greek elements, and his explanation of baseflow is essentially Aristotle's theory of exhalations: that air makes its way into the ground where it is condensed by the cold into water which then supplies the rivers. The air derives its water from blowing over the sea so the idea is partly sound and is consistent with the theory of matter: the rest of the air condenses in the atmosphere to give rainfall which in turns causes floods 'and rivers to flow faster.'

Another subject of great interest was why the sea failed to fill up despite the constant inflow of water from rivers. This problem was doubtless highlighted by the known fact that the Mediterranean was landbound, apart from the Straits of Gibraltar, and even here the surface flow was inwards. Although evaporation was known to the Greeks it was not usually thought to be sufficient to account for the balance. Aristotle preached sufficiency but older sources thought that the water flowed back from the sea to underground reservoirs through tubes, pipes and cavities to provide the base flow of streams and the sources of springs. Seneca's solution is circular:

> The sea is a unity, undoubtedly established just as it was in the beginning. It has its own sources which keep it filled and cause rough waters. As in the case of the sea so also in the case of the gentler waters, there is a vast hidden supply which no river's flow will drain dry. The reason for its strength is obscure. There is emitted from it only as much as it is always able to flow (Book III-14).

Thus Seneca invents a vast hidden reservoir which maintains

equilibrium in the sea and the rivers, a commendable if misguided attempt. But (Book III-15):

> I have this further theory: the idea appeals to me that the earth is governed by nature and is much like the system of our own bodies in which there are both veins (receptacles for blood) and arteries (receptacles for air). In the earth also there are some routes through which water flows and some through which air passes.

Seneca then develops this analogy, especially with respect to blockages and the mutability of matter. The anatomical analogy was to present itself again 1400 years later to Leonardo da Vinci, and 300 years later still, to James Hutton, who however, knew the proper circulation of the body. In all, Seneca is somewhat behind Aristotle and still imbued with a variety of animism.

Seneca, in his youth, wrote a book, now lost, on earthquakes, and this topic was common to almost all Classical authors. The interest is not unnatural in view of the instability of the plate boundaries in the eastern Mediterranean. Detailed descriptions abound of their catastrophic effects in both upraising lands and in sinking them. The classic case is certainly Plato's account, supposedly from the Egyptian Priests, of the destruction of Atlantis. More realistic is Pausanias' account, in his 'Description of Greece' (7.24.5-13), of the virtual disappearance of the town of Helice in 373 BC.

Aristotle attributed earthquakes to wind, produced by dry exhalations within the earth. Seneca reviews a great number of theories of earthquakes based on the four Greek elements, or combinations of them, before deciding 'that the cause of this disaster is air.' The abundance of harrowing descriptions of earthquakes in Classical literature had an undue influence in post-Renaissance Europe when theorists sought powerful and rapid methods of creating topography.

Strabo, a late Greek writer, is the most fitting author with whom to conclude. It is interesting that Lyell (1830) cites Strabo's 'Geography' with considerable approval in his review of Ancient thought. In discussing the distribution of fossil sea shells elevated above sea-level he asserts quite positively (Lyell's translation):

> It is not because the lands covered by seas were originally at different altitudes, that waters have risen, or subsided, or receded from some parts and inundated others. But the reason is that the same land is sometimes raised up and sometimes depressed, so that it either overflows, or returns into its own place again...
> It is proper to derive our explanations from things

which are obvious, and in some measure of daily occurrence, such as deluges, earthquakes, and sudden swellings of the land beneath the sea.

It is easy to see why Lyell, the arch uniformitarian, italicised the last paragraph in his own seminal text. That prescient statement of Strabo's stands sentinel on the pathway to modern understanding, a beacon across the centuries. It is no fault of the Greeks that their more enlightened statements failed to illuminate Renaissance Europe and it is entirely appropriate that the second half of this chapter will examine the failure of another intellectual milieu to overcome the constraints of its cultural environment, at least for several centuries.

DARK AGES, ARAB LORE, SCHOLASTICISM AND LEARNING REVIVED

The decline and fall of the Roman Empire implied too the loss of technology and the decline of learning. By the sixth century Plato's Academy in Athens had been closed, the full texts of Aristotle had been lost and the Greek language itself replaced, first by Syriac, and then by Arabic. The flowing tide of Christianity did little to compensate for this loss: Saint Ambrose complained that men 'lose sight of heaven when they are employed in measuring the earth' and Bishop Theophilus had one branch of the Alexandrian library destroyed in 390 AD. Only in the far north west fringes of Christian Europe, in Ireland, Scotland and the north of England was a modicum of ancient learning preserved and fostered.

At the other end of Europe a slightly later revival was undertaken by the Arabs. Al-Mamun, Caliph of Bagdad, granted peace to the Emperor Michael III in 814 in return for all the writings in the Imperial Library at Constantinople. This Arab resource seeded in turn the revival of learning in Europe in later centuries. The Arabs, in the meanwhile, added something of their own so that although Ibn-Sina (980-1037), also known as Avicenna, adopted many of his views from Aristotle, he made a clear distinction between types of mountains noting that they might arise from two causes:

> either from uplifting of the ground, such as takes place in earthquakes, or from the effects of running water and wind in hollowing out valleys in soft rocks and leaving hard rocks prominent, which has been the effective process in the case of most hills. Such changes must have taken long periods of time, and possibly the mountains are now diminishing in size (in Geikie 1905).

Avicenna probably based his views on 'The discourses of the

29

brothers of the purity' (c941-c982 AD) which describe a geostrophic cycle, including lithification, of remarkable modernity. Regrettably, this work remained untranslated until recently (Said, 1950).

In the same century Omar el Aalem (the Learned) wrote a treatise entitled 'The Retreat of the Sea,' in which he compared charts of his own time with those two thousand years earlier and inferred considerable change along the coastlines of Asia. However, Mohammedan theologians were quite the equal of Christian ones in the matter of adherence to sacred works and Omar went into voluntary banishment rather than admit that certain passages in his works were contrary to the Koran.

At about the same time (1080 AD) Shen Kua in China was inferring climatic change from bamboo shoots buried in river alluvium, shifts in the position of the sea from fossil shells and coastal deposition from the appearance of the sediment-laden Hwang Ho river. The Chinese had already invented a seismograph in the second century AD and had earthquake records dating back to 780 BC. Regrettably, although fragments of their science and technology filtered through Asia to Europe over the centuries, such as silk cultivation to the Ancient World and the compass to the late Medieval World, none of their abstract science became known in the west until it was too late to have a direct influence.

The revival of learning in Europe began about 1250. Between 1200 and 1225 Aristotle's full texts were recovered, translated and discussed. The reconciliation then achieved by Thomas Aquinas (1224-1274) between sacred and profane learning became known as Scholasticism, and while it taught an appeal to reason it did not stoop to the experimentation preached by Roger Bacon (1214-1294); a true man of science trapped in a medieval mind (Dampier 1966). As usual, the hypothesis of a master became the dogma of disciples, and Aquinas' adoption of Ptolemaic astronomy rapidly became an article of Christian faith that proved difficult to dislodge. More generally, by 1500, Classical sources were so welded to to Christian faith that an attack on the former was construed as a denial of the latter, despite some rather blatant discrepancies, especially in the matter of chronologies.

THE ITALIAN RENAISSANCE

With the invention of printing in the mid-fifteenth century, its rapid diffusion throughout Europe before 1500, and the re-discovery of America in 1492, men's minds were ready, both inwardly and outwardly, to explore the wonders of the universe. Columbus' voyage was, in itself, a metaphor for the forthcoming age: a navigational experiment on global geography - for such was the design, if not the result.

For scientists it is usual to see Leonardo da Vinci (1452-1519) as the harbinger of the modern era, but it is a recognition tempered by the fact that much of his most valuable work remained secret until his encoded notebooks were transcribed and made public centuries later. A recent study of da Vinci's fluvial notes (Alexander 1982) was forced to conclude that 'his studies of natural landscapes were not seminal but mark the beginning of the transition from purely theoretical to observational and deductive methods of analysis in earth science.' Leonardo's secrecy may have stemmed from the plainly heretical nature of many of his conclusions in which it is clear that he had a firm grasp of erosional principles from what would now be termed a uniformitarian viewpoint. Most of his fluvial analyses were of a hydraulic character and were motivated by practical desires to improve public works on rivers and canals. He realised that 'a river which has to be diverted from one place to another ought to be coaxed and not coerced with violence,' and it is typical of his engineering attitude that he actively sought the manuscripts of Archimedes. Curiously, despite his deep understanding of hydraulics, which needed almost two centuries of work by subsequent Italian engineers to equal, he misunderstood the true nature of the hydrological cycle. His views, though similar to those of Seneca and Pliny, probably came from the Roman engineer Vitruvius (Alexander 1982), and they were, perhaps, unduly influenced by his anatomical studies, although based on very thorough investigations and much thought. His premise was that 'we must begin with experiment' and if, at times, he looked over his shoulder to Classical engineers he simultaneously foreshadowed much in the centuries ahead. He knew that water was 'the vital humour of the terrestial machine' three centuries before Hutton, but his period placed a stranglehold on his science, while approving and financing his art and military mechanics.

The onset of the Renaissance did not bring immediate benefits to what would now be termed science; it merely provided a different nursery for the propagation of ideas. The Classical authors became better known through the medium of the printed word and they remained a ready and trusted source. For example, Caxton's encyclopaedia 'The Myrrour of the Worlde' (1481) contains a passage suspiciously like that from Seneca quoted earlier:

> Just as the blood of a man runs through the veins of the body and issues forth at a particular place, so the water runs through the veins of the earth and comes to the surface in springs and fountains.

The failure of the Classical world to develop a physics that could be translated into mechanical principles via

31

mathematics was only gradually rectified as the Renaissance progressed. Da Vinci had stressed the need to experiment but the lesson was not properly heeded for another century when Francis Bacon (1561-1626), in his turn, preached the need to record and tabulate all possible observations and experiments. Galileo Galilei (1564-1642) made very significant advances in mechanics, and practised what Bacon preached, but his confrontation with the Vatican and the threat of the Inquisition put an end to the scientific tradition in the Mediterranean, as Galileo himself forecast in 1630. Descartes (1596-1650) likewise judged it wiser to suppress his book 'Traite du Monde' on hearing of Galileo's experience. Newton (1642-1727) finally formalized and synthesized the mathematics, mechanics and methodology needed to operate the physics of the ordinary world, and it is significant that his 'Principia' were quite timeless and preached no theory of the universe, or of the earth. Newton's theological views were kept quite separate from his physics. From the point of view of the earth sciences, however, mechanics and experimentation was of no great usefulness when theoretical schemes of geology, contemporary with Newton, involved the whole globe and even attracted his tacit approval, and when observation in the field had yet to equal that of the better Classical authors.

POST-RENAISSANCE EARTH SCIENCE: 1578-c1800

I characterized the Classical period as one lacking a progression of ideas. The post-Renaissance period is more difficult to label. Davies (1969), in his masterly study of British geomorphology, divined a clear periodicity with periods of fundamentally sound observations giving rise to increasingly refined, speculative, and fantastical theories. The intellectual poverty of these theories caused a relapse before renewed observation generated a new theoretical momentum. Davies noted peaks of theoretical activity at the close of the seventeenth, eighteenth and nineteenth centuries. The picture on the Continent was more complex. French and Italian authors showed an erratic progression of ideas with French efforts climaxing toward the end of the eighteenth century, but with Italian writers petering out after the writings of Moro and his illustrator Generelli. German workers appeared to specialize more in mineralogy and the subject of fossils until the rise of the great German travellers at the close of the eighteenth century. Space prohibits a detailed exposition of all the many and varied theories put forward over two centuries, and for which Lyell (1830 and all later editions) and Zittel (1901) still provide good reviews, and I intend to discuss the period under a number of different themes, with only incidental reference to national schools of thought. A number of problems beset those who developed an interest in the history of the earth and most of these stemmed from a professed belief in the Bible. In practice belief was selective with strict adherence at a few crucial points such as the Creation and the Flood.

Background beliefs and scriptural authority

The overwhelming opinion in sixteenth and seventeenth century Europe was that nature was degenerate and that the deficiences in the world could be traced to man's earthly sins. Notwithstanding this, God's wisdom stood revealed in the world as it was seen. The Bible was taken to be the ultimate authority and those who denied it risked social ostracism, or even their lives. These fears relaxed to some extent as the two centuries passed: Giordano Bruno was burnt at the stake in 1600 for his heterodox views on earth history, including the efficacy of running water, a fate that would have been inconceivable anywhere in Europe by 1700. Nevertheless, a failure to conform brought other perils for intellectuals. In England, Burnet (1635-1715), who found Court favour with his 'The Theory of the Earth' (1684), lost it with a second edition (1690) in which he was emboldened to suggest that the Book of Genesis was allegorical, not literal. William Whiston (1666-1752), selected to follow Newton at Cambridge, was expelled from his chair in 1710 for his heterodoxy, and Edmond Halley (1656-1742) thought it better not to publish several papers on the Flood that he had presented to the Royal Society.

The situation was but little different on the Continent. Leibnitz (1646-1716), according to Zittel (1901), phrased his famous 'Protogaea' (written 1680, published 1749) with an ambiguity consistent with personal safety. Buffon (1707-1788), as late as 1751, was asked by the French Academy to recant various sections of his 'Histoire Naturale' and to publish a retraction in his next work. This he did with respect to 'everything in my book respecting the formation of the earth.' By 1778, when he published 'Epoques de la Nature,' matters had improved to the extent that, although far more revolutionary (it calculated an age for the earth of 93,000 years), it did not attract the same opprobium. De Maillet (1656-1738), who advocated an eternal Earth and gradual changes of sea level, took the precaution of publishing his work 'Telliamed' (1749, English translation 1750), written about 1719, eleven years after his death! In Italy, Moro (1687-1740), an Abbot, was careful to obtain an official affidavit for his preface to the effect that the book contained nothing inimical to the Catholic faith. It can hardly be doubted then that bibliolatry was a powerful force shaping and shadowing every thought that might try to rise unforced from field observations. Curiously, scriptural

authority appears to have been strongest in Protestant
countries, if only because the rejection of papal discipline
meant its replacement by the literal word of the Bible.

Biblical chronology and 'The Complaint of Fishes'

Two particular problems sprang from a literal belief in the
Bible: a short earth history, and the Mosaic Deluge or Flood.
The most crucial was certainly the extremely short timescale
that the Bible implied for the operation of processes that
shaped the surface of the globe. A brief timescale was quite
contrary to the concept of the eternal universe as taught by
Aristotle, and almost all Classical authors. Prior to the
middle of the seventeenth century the precise time available
since Creation was presumably limited but usually
unspecified, so that William Bourne, an innkeeper, could
write in 1578, that 'the age of the worlde is no small tyme'
without it being clear whether he accepted the biblical
chronology, or was proposing a more extended one. The ninth
century Anglo-Saxon Chronicle gives a date of 5193 BC for the
Creation, but, in 1654, Bishop Ussher published his considered view that the Creation took place on 23 October 4004 BC
and the Flood from 7th December 2349 BC to 6 May 2348 BC. The
Creation date was incorporated into the King James edition of
the Bible in 1701, and as Davies (1969) has remarked,
Ussher's reputation as a savant ensured widespread acceptance
for the date. A compounding problem was that, in the
sixteenth and seventeenth centuries, the end of the world was
both forecast and expected: a common date, until disproved,
was 1657.

The other inescapable problem was the Flood story
recorded in the Bible. Here, the Classical authors were in
partial agreement with the Scriptures: Plato and Aristotle
both mentioned extensive floods, of Atlantis and Hellas
respectively, so the two sources seemed to support each other
in kind, if not in details.

The Flood story was bolstered by the abundant evidence
of fossils. Both in antiquity, and in the Renaissance, there
was extensive discussion of the vexing question about the
true origin of fossils, primarily marine shells. Opinions
varied from the (modern) truth, to the idea from Aristotle
and others, that there was 'plastic virtue' in the earth
which caused them to be formed, as 'sports of nature' or as
imitations of living forms. As an acceptance of their real
nature became the majority, but by no means unanimous view,
so the Flood was seen as a suitably documented mechanism for
their emplacement. The Flood also served as a convenient
agent for the shaping of the major landforms of the earth,
and it was a notion that persisted until late in the
nineteenth century, in one form or another. The hydraulic
properties ascribed to the Flood were quite as variable as

the imaginations of the authors and, for a period so prepared to accept the Bible as a source on earth history, the apparently tranquil and short-lived nature of the Flood recorded in Genesis was widely ignored. Apt indeed was the title of Scheuchzer's work (1708) that discussed the remodelling of the earth by the Flood: 'The Complaint of Fishes.'

The reality of denudation

The delicate dependency of transportation on landscape form and process that existed prior to the nineteenth century made some appreciation of denudation inevitable amongst travellers, whatever their station in life. The more important issue was to appreciate the scale upon which it operated and its implications in the long run. From the opening of the Renaissance to the beginning of the nineteenth century there were always some writers in Europe with a firm grasp of erosional principles, both in the short run - daily decay, and in the long run - earthly decay.

After da Vinci, Palissy (1510-1590) and Bauer, better known as Agricola (1494-1555), both gave extensive discussions of denudation. For example, Agricola (1656):

> We can plainly see that a great abundance of water produces mountains, for the torrents first of all wash out the soft earth, next tear away the harder earth, and then roll down the rocks, and thus in a few years excavate the plains or the slopes to a considerable depth... By such excavation to a great depth through many ages there rises an immense eminence on each side. When an eminence has thus originated, the earth slips down loosened by constant rain and split away by frost, and the rocks, unless they are extremely firm, since their seams are similarly softened by the damp, roll down into the excavations below. This continues until the steep eminence is changed into a slope...Moreover, streams, and, to a far greater extent, rivers, effect the same results by their rushing and washing... By a similar process the impetus of water completely demolishes and flattens out hills and mountains (Book 3, p1-2).

In England, William Bourne (1535-1582), innkeeper and gunner at Gravesend on Thames-side, is given pride of place by Davies (1969) for his recognition of the efficacy of ordinary erosion. His book (1578) was intended for travellers, but the fifth part of it describes 'the naturall causes of Sands in the Sea and rivers, and the cause of marish ground, and Cliffes by the sea Coasts and rockes in the Sea.' His passage on sea cliffs is especially enlightening:

> My opinion is thys, as the age of the worlde is of no
> small tyme, so in process of tyme the often sufferynges
> of the bellowes of the Seas have beaten away the feete
> of those hilles, that are by the sea coastes. And so
> undermyning it, although it were of harde stone, yet the
> wayght of that which was undermined hanging over, in
> rayny wether, or after great frost, must needes fall
> downe into the Sea. And then that soyle or substaunce
> that fell down, in process of time was beaten or washed
> away agayne, by the often soussing of the bellowes of
> the sea, in the time of great wyndes and stormes. And
> then the stuffe so fallen down, being washed and
> consumed away, the sea doth begin to undermine it
> agayne, by litle and lytle (quoted in Davies 1969).

Bourne goes on to describe the abrasion and rounding of
pebbles on the beach and clearly he had a perfectly modern
view of coastal processes insofar as he described them.

Subsequent writers such as Nathanael Carpenter (1625)
and Bernhard Varenius (1672) provided general discussions of
denudation, the former quite comprehensively, and the latter
claiming that the Hwango Ho was a third part sand and gravel.
Nicolaus Steno (1638-1687), whose works were translated into
English by Henry Oldenburg (1671) of the newly-formed Royal
Society, made denudation a part of his stratigraphical
account of the plain of Tuscany. Even the clerical theorists
of the late seventeenth century, (Burnet 1690, 1691, Bentley
1693), did not deny denudation. The momentum was carried into
the next century by the posthumous works of Robert Hooke
(1705), the writings of John Ray (1673, 1693, 1718) and
reprinted editions of Burnet and their contemporaries. A
fluvial school was perpetuated in France by the writings of
Gautier (1721), Guettard (1752), Desmarest (1775), Montlosier
(1778), and finally Lamarck (1802).

Curiously, the short biblical timescale did nothing to
inhibit a belief in natural denudation, but the inevitable
conflict between theory and practice had to be resolved in
some way and this I will discuss under the heading the
'denudation dilemma', after a brief review of seventeenth-
century theorists.

Seventeenth-century theorists

The late seventeenth century was notable for the theoretical
works of several writers, particularly English ones whose
work was frequently reprinted and translated into French and
Italian. However, pride of place is often given, but probably
erroneously (Drake and Komar 1981), to the work of Nicolaus
Steno (1669), a Dane living in Italy, who attempted an
interpretation of a specific landscape, that of Tuscany, and
who recognized a series of events organized into a crude

stratigraphic succession. The work was quickly translated into English and must have been known to Hooke (1635-1703), curator of experiments at the newly-formed Royal Society. Steno attempted to harmonize his interpretation with the scriptures, yet he found it necessary to call upon two periods of crustal collapse to explain the landscapes he saw. His ideas are summarized in the diagram already given in Chapter One (Figure 1). The idea of subterranean caverns was, of course, consistent with classical writers and not unsupported by field evidence in an area rich in limestone. It was also a feature of a work of immense popularity in seventeenth-century Europe: Kircher's 'Mundus Subterraneus' (1664). This work had little influence in Britain, although it illustrated profusely the favourite classical idea that spring and river waters were conveyed to the tops of mountains by underground channels, the earth's structure being similar to a Swiss cheese!

Steno's contemporary, Hooke, had very advanced ideas on geology but unfortunately his 'Discourses,' based on notes and lectures given in the period 1668-1699, were not made public until 1705, after his death. Some knowledge of them was clearly available to his scientific contemporaries in the Royal Society, and some of his experiments involved exploring the mechanisms responsible for the formation of lunar craters, but his ideas did not figure in the public debates surrounding the popular works by Burnet, Whiston, Woodward and others. Davies (1969) notes that Hooke had 'a shadowy outline of a theory of the Earth that is almost identical with the renowned theory which Hutton presented .. in 1785,' and his French contemporary Gautier (1721) had similar views involving prolonged erosion and periodic uplift.

More typical of their age, and more popular with the public, were the works by Burnet, Whiston and Woodward. Burnet's 'The Theory of the Earth' (1684,1690) envisaged a featureless pre-diluvial globe that was submerged by flood waters from within the earth, which then left the 'world as a gigantic hideous ruin': the visitation of God's wrath upon man. This puritancial view left Burnet free to recognize the effects of denudation, the reality of which he freely admitted. The theory, which Buffon later termed 'a fine historical romance,' attracted fierce criticism from clerics on the one hand and mathematicians on the other but in its first edition won Burnet preferment at Court. Whiston (1696), in 'A New Theory of the Earth,' adapted Burnet's scheme but stimulated by the recently discovered Halley's comet, used a comet to disturb the ocean and create the Flood.

The third writer in this trio was John Woodward (1665-1728) who had at least his feet on the ground:

      I made strict enquiry wherever I came, and laid out for intelligence of all Places where the Entrails of the

Earth were laid open, either by Nature (if I may say so) or by Art, and human Industry. And wheresoever I had notice of any considerable natural Spelunca or Grotto; any digging for Wells of water, or for earths, Clays, Marle, Sand, Gravels, Chalk, Cole, Stone, Marble, Ores of Metal, or the like; I forthwith had recourse thereunto: and taking a just account of every observable Circumstance of the Earth, Stone, Metal, or other Matter, from the surface quite down to the bottom of the Pit, I entered it carefully into a journal, which I carry'd along with me for that purpose. ... and 'tis out of these Notes that my Observations are compiled.

These admirable sentiments were buttressed by a questionnaire administered to people abroad from which he convinced himself that conditions elsewhere were essentially as they were in Britain. Apart from these exploits in the field, Woodward was also notable for his different, Restoration, theological attitude to the Flood which in contrast to Burnet's view he saw as 'a most monumental Proof..of Goodness, Compassion and Tenderness.' However, his vision of the Flood as a hydraulic agent was no improvement on Burnets'. He conceived 'the whole terrestial globe to have been taken to pieces and dissolved at the flood, and the strata to have settled down from this promiscuous mass as any earthly sediment from a fluid.' Regrettably then, his field work, though springing from his belief that 'Observation (was) the only sure grounds whereon to build a lasting and substantive philosophy,' was of no avail in his theory construction. Woodward endowed and named the chair of Geology at Cambridge although, with the exception of a brief tenure by Michell, who wrote a comprehensive and valuable account of earthquakes (1760), it had no significant occupant until Sedgwick in 1818.

. Woodward's view of denudation was ambiguous and coloured by his theology which held that the post-diluvial globe was the creation of God, and not liable, therefore, to be in a state of serious disrepair. Mild surface erosion he admitted, but coastal change he emphatically denied (Davies 1969).

The opposite view was held by John Ray (1627-1705) who invented stories which exaggerated the rate of surficial erosion to the extent that a hill in Derbyshire was sinking visibly, within the lifetime of men! Ray's problem was that the observed rate of erosion was slower than that required to accomplish what could be seen to have occurred in the six thousand years since Creation. Ray's discourses, however, held none of the imaginative charm of the previous three, and a critical public preferred biblical geo-fantasy.

The new belief that the world was a perfect abode created for man, in combination with obvious surficial denudation, led to the eighteenth-century problem known as the 'denudation dilemma.'

The Denudation Dilemma

If mountains are a necessary part of the terrestial economy, as created by God in His wisdom, is it a part of the divine scheme that denudation should lead to their decay? Prior to 1700 the majority view was that mountains were ugly excrescences: warts, boils and blisters manifesting the sins of man. That these, and the world in general, were in a state of rapid decay was therefore no surprise. The perfect globe, if it existed, had done so before the Flood, which Genesis states had been sent by God in His wrath. In addition, since the end of the world was expected so soon, the effects of denudation were in any case irrelevant.

By 1700 the attitudes had all changed. The world was no longer expected to end with such certainty, denudation was an accepted reality, yet the world was supposed to be perfect. The resolution of this problem was by one of three means. The first solution was to argue in favour of a process of repair or restoration taking place below the ground. The idea of rock growth goes back in part to the Ancient World but was certainly extant in the mining communities; perhaps based on the observation that unsupported tunnels eventually close up by a process of creep. As late as 1838, Bakewell was cautioning his readers against the notion that rocks could grow, organically, from seeds!

The second solution was to argue that mild erosion had a cleansing effect on the landscape and provided fertilized soil for the valleys below the mountains. In this way denudation was beneficial to man, the globe became even more habitable, and the mountains slowly decayed to a more useful state. Some writers perceived that the long term results of this would be a uniform surface, but did not anticipate that the debris would be carried into the oceans. Williams (1789) expresses these sentiments about debris which is:

> all well and wisely disposed of for the benefit and advantage of the present earth, and inhabitants of it. Part of it is lodged in lakes and unseemly gulphs, in the course of rivers, which are improved thereby into rich and pleasant valleys and plains, and the residue is carried along by the floods, to the borders of the oceans, where it is very happily disposed of to form new land, which in fact enlarges the bounds of our habitations, and in time becomes the most useful, the richest, and most convenient parts of the earth for society and commerce.

The final solution was to deny erosion altogether: the extreme end of the spectrum, with Williams occupying a moderate position. The actuality of some present denudation was admitted, but that it could go on forever was denied.

Woodward had already adopted all three resolutions of the dilemma and argued that debris at the coast was prevented from sinking by the 'greater crassitude and gravity' of sea water, and that some of it was returned to the land by way of evaporation! In his anxiety to berate his contemporaries for their reliance on ancient sources he himself denied Aristotle and Herodotus while providing a dull echo of one of Pliny's less accurate 'facts': 'All seas excrete refuse at high tide, some also periodically.'

IN CONCLUSION

These attitudes held sway for almost a century and even lingered into the early nineteenth century, yet despite this a belief in denudation did persist with moderate observers, as common sense would suggest. Italian and French writers maintained a belief in fluvial erosion throughout the eighteenth century. Guettard (1752) in particular recognized the importance of surficial denudation in explaining the volcanic topography of the Massif Central; a curious yet potent juxtaposition of the elements fire and water, and a theme that would recur in the writings of Desmarest (1775) and Montlosier (1788) although the main geological controversy at this time was the origin of volcanic rocks, not the work of rivers. Italian naturalists also described the ability of rivers to transport sediment and to excavate their valleys (Targioni-Tozzetti 1752), although they also emphasized the importance of the changing level of the sea as recorded by recent marine fossils (Moro 1740, Generelli 1749). The writings of the French school subsequently exerted a significant influence on Playfair, Scrope, and Lyell.

To some extent the temporal shackles were already being loosened: allegorical interpretations of the 'days' of Creation began to appear as 1800 approached. A resurgence of theory building began with a greater emphasis on the details of local topography, and the potential impact of Flood waters or earthquakes thereon. In some respects these were to be throwbacks, especially in Britain, to the bizarre schemes just reviewed. But already a flanking movement was abroad, and at least one individual had been stalking the sideroads of Britain and the Continent, hammer in hand, and prepared to favour field evidence over the literal text of the Bible.

Charles Lyell (1830) forewarned his readers that his review of the eighteenth century would 'be occupied with accounts of retardation, as well as the advance of science ... the frequent revival of exploded errors, and the relapse from sound to the most absurd opinions.' This assessment still holds true and in consequence the theories arising in the late eighteenth century are best evaluated in the light of the one of them that survived.

Chapter Three

HUTTON AND PLAYFAIR VERSUS THE REST: NASCENT GEOLOGY

> *"It is snowing still,"* said Eeyore gloomily.
> *"So it is."*
> *"And freezing."*
> *"Is it?"*
> *"Yes,"* said Eeyore. *"However,"* he said, brightening up a little,
> *"we haven't had an earthquake lately."*
>
> (A.A. Milne, The House at Pooh Corner)

England in the period after 1727 basked in the reflected glow of Newtonian mathematics and physics, but did little to augment it. In contrast, the French and Swiss developed an upsurge, inherited the intellectual mantle of Newton, and gave to science such great names as Laplace, Lagrange, Buffon, the Bernoullis, Euler and Bouguer. As a counter-vailing influence, and even more remote than Paris in actual travel time, the Scottish Universities, and particularly Edinburgh, also developed a healthy spirit of intellectual freedom, producing for example Colin Maclaurin, David Hume, Adam Smith, Joseph Black and James Watt.

The previous chapter documented the existence of the denudation dilemma in eighteenth-century England as an intellectual conflict between the notion of a divinely perfect globe of recent origin and the obvious existence of denudation. The emphasis was always on the conflict between perfection and decay, rather than on the limited time since Creation, per se. Since, in hindsight, we know that denud-ation is real, then clearly it was the Mosaic record which had to be jettisoned. The problem was two-fold: on the one hand denudation as an explanation of landforms necessarily required a timescale that was orders of magnitude larger than the biblical one, and on the other hand any attempt to interpret the geological record in terms of the biblical record, with an inevitable emphasis on the Flood, was doomed to failure, whatever timescale was adopted.

Progress depended on an ability to reject both aspects of biblical natural history, but loosening the shackles of time was easier than rejecting a sacred narrative. As early as 1695 Edward Lhwyd noted that the present rate of fall of boulders into the passes of Snowdonia was quite insufficient to account for the number observed and hinted in a letter to John Ray that 'in the ordinary Course of Nature we shall be compelled to allow the rest many thousands of Years more than the Age of the World.' Various writers, from Leibnitz and Burnet to Ray and Buffon had speculated that the days of

Creation where not sidereal days, and so drew overdrafts on what was imagined to be the holy bank of time. But while such imaginary raids provided time to create the world's topography they were no insurance against the evidence of long historical chronologies in Egypt, Babylon and China, or against the immense timescales that a careful study of stratigraphy would reveal. Woodward's exploits in the field only served to reinforce his preconceived theory, although they were steps in the right direction. The alternative concept of an eternal world had always been available in the works of Classical authors, and this idea was revived during the eighteenth century, particularly after the lapse of the Press Licensing Act in 1695 (Davies 1969). Believers in the revelation of God through his works of nature, Deists, usually adopted an Aristotleian view of the universe.

It is was into a changeable, less pre-ordained, world that James Hutton was born in Edinburgh, the year before Newton's death.

JAMES HUTTON (1726-1797)

Most of what is known of the life of Hutton is contained in a biographical account given by his friend and associate John Playfair (Playfair 1805). The son of a wealthy Edinburgh merchant and sometime City Treasurer, he attended the city High School and entered University in 1740 (aged 14!) to study humanities. Inspired by a chance acquaintance with chemistry, he quickly left an apprenticeship in the Office of the Signet, and re-entered the University to study medicine from 1744-7. At this point, aged twenty, he left to complete his training in Paris, finishing it in Leyden where he gained his M.D. with a dissertation on 'Of Blood and the Circulation of the Microcosm.' He returned to London and a reluctance to return immediately to Edinburgh to practise as a Doctor is conceivably linked to the fact he had fathered a son (but remained a lifelong bachelor) in about 1747 (Robinson and McKie 1970, see letter 184) and which Donovan and Prentiss (1980) suggest may have created a local or family scandal. If so, the scandal was hushed up successfully since no hint of it was known to his subsequent Edinburgh colleagues, or apparently his surviving sister Isabella, until James (jr) appeared in Edinburgh on his father's death in 1797.

Instead of medical practice, Hutton decided to pursue an interest in agriculture on a small estate he had inherited at Sligh Houses, Berwickshire, about 40 miles from Edinburgh. To educate himself in agriculture, he spent two years in Norfolk (1752-54), during which time he made walking tours through much of southern England, and made a further visit to the Low Countries before finally settling in Berwickshire, with a ploughman imported from Norfolk.

It was probably during this agricultural apprenticeship that Hutton developed his geological interests. He had already made the acquaintance of Sir John Hall, a landowner living near Sligh Houses, and in 1753 he wrote to say he had become 'very fond of studying the surface of the earth, and was looking with anxious curiosity into every pit, or ditch, or bed of a river that fell his way' (Playfair 1805). It is clear from later descriptions in his published works that he had travelled far and studied well.

The combination of chemistry, agriculture and a thorough medical training at the best centres in Europe appears to have been a suitable mixture for a budding geologist. In addition to these qualities, Hutton brought an instinctively logical mind to bear upon all the problems confronting him. The crucial factor, however, appears to have been his intimate acquaintance with field evidence, a point that has possibly been understated in contrast to the partially deserved criticism of his convoluted prose.

Hutton possessed the essential requirements for making theoretical progress in geology, and these qualities were present as early as his rather rushed and hesitant Leyden thesis. Donovan and Prentiss (1980) remark that Hutton brought to this thesis a methodological approach based on a chemical philosophy nutured in Edinburgh and Paris, and borne too of personal predilection and experience gained before he left Edinburgh when he and a fellow student (John Davie) devised a method for extracting sal ammoniac (ammonium chloride) from soot (Clow and Clow 1947). Commercial production of this popular agricultural fertiliser provided a sound financial basis for Hutton's life while his chemical philosophy suggested to him that 'nature is ordered in a certain way and that that order can be best known through certain kinds of experiment' (Donovan and Prentiss 1982).

This ability to reason successfully about observations was kept carefully separate from a belief in God whom he recognized in 'the glorious cycle of life and a very beautiful instance of perpetual moving' (Hutton 1749). The doctrine of the macrocosm and the microcosm: that life in the body was a minature replica of the global organism and that nature is made up of a multitude of closed cycles, was not seriously pursued in the dissertation. In subsequent writings it, and a design motif, permeate the prose as a motivational logic underlying the whole without seriously undermining the physical and chemical logic of the steps enunciated in the various versions of his 'Theory of the Earth.' An unpublished preface to his Theory, recently discovered (Dean 1975), is a perfect expression of Hutton's deistic beliefs. In it, Hutton tries to forestall religious criticism, and uses the old argument that the days of creation cannot be sidereal days. He concludes that it is impious to doubt either the Scriptures, or the findings of natural philosophy!

'Natural History of the Earth'

The earliest geological sketch that Hutton is known to have made was 'written at a very early period' (Playfair 1805) and probably dates to about 1760 or just before. It is worth quoting Playfair's account in full since it contains the essence of Hutton's subsequent work:

> From these sketches it appears that the first of the propositions just enumerated, viz. that a vast proportion of the present rocks is composed of materials afforded by the destruction of bodies, animal, vegetable, and mineral, of more ancient formation, was the first conclusion that he drew from his observations.
>
> The second seems to have been, that all the present rocks are without exception going to decay, and their materials descending to the ocean. These two propositions, which are the extreme points, as it were, of his system, appear, as to the order in which they became known, to have preceded all the rest. They were neither of them, even at that time, entirely new propositions, though, in the conduct of the investigation, and in the use made of them, a great deal of originality was displayed. The comparison of them naturally suggested to a mind not fettered by predjudice, nor swayed by authority, that they are two steps in the same progression; and that, as the present continents are composed from the waste of more ancient land, so, from the destruction of them, future contin- ents may be destined to arise. Dr. Hutton accordingly, in the notes to which I allude, insists much on the perfect agreement of the structure of the beds of grit or sandstone, with that of the banks of unconsolidated land now formed on our shores, and shews that these bodies differ from one another in nothing but their compactness and induration.
>
> In generalizing these appearances, he proceeded a step farther, considering this succession of continents as not confined to one or two examples, but as indefinitely extended, and the consequence of laws perpetually acting. Thus he arrived at the new and sublime conclusion, which represents nature as having provided for a constant succession of land on the surface of the earth according to a plan having no natural termination, but calculated to endure as long as those beneficent purposes, for which the whole is destined, shall continue to exist.
>
> This conclusion, however, was but a suggestion, till the mechanism was inquired into by which loose materials can be converted into stone, and elevated into land. This led to an investigation of the mineralizing

principle, or cause of the consolidation of mineral bodies: And Dr Hutton appears accordingly, with great impartiality, and with no physical hypothesis whatever in his mind, to have begun with inquiring into the nature of the fluidity which so many mineral substances seem to have possessed previous to the acquisition of their present form.

This extended extract reveals Hutton as field worker and theoretician, and from the point of view of geomorphology what remained for the future was merely elaboration, important certainly, but still essentially decoration of an 'otherwise bald and unconvincing narrative.' The logical need for repeated phases of continental creation (perhaps suggested by 'puddingstone' i.e. conglomerate), for 'laws acting perpetually' and the necessity for indefinitely long 'geological' time are all present in this early sketch. Notably lacking are catastrophic overtones, earthquakes or floods, or any reference to biblical schema: it is all modern in tone, although admittedly in Playfair's prose.

The essential geological problem remained: how to indurate sediment and upraise it. The exact mechanism eventually used by Hutton, compression and heat coupled with latter's expansive properties, is of only secondary interest to geomorphology. The idea of uplift, was however, important, since it provided the conceptual possibility of relative changes of land and sea levels: a conundrum that is still not completely understood, and which has provided intimate ties between geophysics and geomorphology.

By using the title 'Natural History' for this sketch, Hutton reveals that in contrast to later writings when the term System or Theory is used, he was aware that he lacked a full theoretical cycle, and had so far only speculated on the mechanism of renovation: a nice distinction. There may be little point in searching the literature for sources that Hutton consulted. Playfair tells us that of 'voyages, travels, and books relating to the natural history of the earth, he had an extensive knowledge,' collating, comparing, and correcting them but 'on the other hand he bestowed but little attention on books of opinion and theory...He was but little disposed to concede anything to mere authority; and to his indifference about the opinions of former theorists, it is probable that his own speculations owned some part, both of their excellencies, and their defects.'

Given the theoretical sources available around 1760, Hutton's attitude was certainly understandable; only Hooke (1705) and Gautier (1721) were worth a second look, and it has been suggested (Drake 1981) that in his later writings his tone is most polemical when he attacks the (unquoted) opinions of Hooke, especially on the issue of a change in the axial tilt of the earth. The issue cannot be resolved with finality although on Playfair's death (1819) Hooke's work was listed in his library. In view of Playfair's remarks, Gautier's title, 'Nouvelles conjectures sur le Globe de la Terre,' probably would not have attracted Hutton.

The only two contemporary works of any significance were both typically cataclysmic. Michell's work on earthquakes (1760) was a useful inventory, stimulated by the Lisbon earthquakes, and its emphasis on caverns and fissures is probably related to a knowledge of the limestone terrain of his native Yorkshire. Catcott, a Bristol cleric, in his 'Treatise on the Deluge' (1761), was concerned to refute allegorical interpetations of the Mosaic Deluge, and attempted to show that the Deluge was both literally true, and universal as well. He emulated Woodward, and like him engaged in field work with a view to validating, rather than testing, his theoretical opinions.

'The System of the Earth, its Duration, and Stability'

The sketch just outlined is difficult to date with any precision but it was probably prepared by 1760 at latest since its observational contents could easily have been gathered by such an assiduous traveller and perceptive farmer as Hutton. The completion of the Theory may have taken another ten years, devoted as Hutton was to mineralogical investigations and collections. In 1764 he made a lengthy excursion through northern Scotland and saw granite in situ for the first time. He moved his residence to Edinburgh about 1768, and became acquainted with Joseph Black the chemist, and Adam Smith the economist. From that time on, Hutton appears to have devoted his life to scientific pursuits.

One of Black's closest friends was James Watt, the engineer, who moved to Birmingham, in the English Midlands, about 1773. Hutton admired Watt's 'reciprocating engine' in 1774 while on a visit to Birmingham, from whence he made a trip into Wales designed to reveal the origin of the abundant 'indurated gravels of granulated quartz' found in the English Midlands. The problem of the gravels was important because, as Hutton realized, an explanation of surficial decay and transport by an appeal to entirely fluvial processes required an explanation of what we now recognize as glacial drift, usually either till or fluvio-glacial deposits. This was a theme which was to remain troublesome until the complete adoption of the glacial theory about 1860, and Hutton clearly felt that it had to be formally accommodated within his Theory. It was, of course, the identical deposit that appeared to offer such convincing evidence of a catastrophic deluge to those yet unskilled in sedimentology.

Hutton convinced himself that the origin of the quartz gravel was an outcrop of strata between Bromsgrove and Birmingham, although Playfair subsequently thought the source was via the medium of what we now call the Bunter Sandstone (Playfair 1802 p375, 1805 p48). Having found the source, Hutton was content that the gravel could be understood as a residual gravel, partly isolated by general fluvial lowering, or, as he would have termed it, 'decay' of the landscape.

The title of this section is that of the abstract prepared for the first public presentation of the Theory in 1785, shortly after the founding of the Royal Society of Edinburgh in 1873. The paper was read in two parts in January and March, and the abstract, required by the Society's regulations, was distributed abroad in that year to friends of the Edinburgh savants, but it drew little critical fire. A facsimile is reproduced in Albritton (1975). Bailey (1967) has disputed Hutton's authorship of the abstract, established by Eyles (1948, 1955), but I believe (Tinkler 1983a) that all his objections, primarily textual and stylistic, can be answered: in particular the abstract is permeated by Hutton's underlying appeal to the principle of 'design' in nature. In Hutton's undoubted words, which conclude the second volume of his 'Theory of the Earth':

> A system is thus formed, in generalising all those different effects, or in ascribing all those particular operations to a general end. This end, the subject of our understanding, is then to be considered as an object of design; and, in this design, we may perceive, either wisdom, so far as the ends and means are properly adapted, or benevolence, so far as that system is contrived for the benefit of beings capable of suffering pain and pleasure, and of judging good and evil.

I shall not review the abstract in detail since it differs from the previous 'Natural History' only by virtue of the greater attention paid to the methods of mineralisation, consolidation and uplift. These are topics whose details are of no immediate interest here, although they are of great importance to geology as a whole since they signify the completion of the Theory in Hutton's mind and establish quite clearly his version of the geostrophic cycle. The one conclusion which should be borne in mind is the generally dynamic type of cyclic steady state that Hutton appears to have had in mind. Unlike almost all other existing 'Theories' at that time, Hutton was not tied to a specific temporal framework and the very generality of his ideas, from which all the conceptual power flows, must have left his average reader floundering - like a whale stranded by a more familiar Mosaic Deluge.

As an example, when the issue of geological time is discussed, Hutton concludes that:

> as there is not in human observation proper means for

> measuring the waste of land upon the globe, it is hence inferred, that we cannot estimate the duration of what we see at present, nor calculate the period at which it had begun; so that with respect to human observation, this world has neither a beginning nor an end.

Hutton was rather pessimistic here, although it might be that he preferred to reserve judgment on the few suggestions that had been made in the contemporary literature on this issue, if they were in fact known to him. The vague generality, that the exact period of time involved is, relatively speaking, immaterial to the discussion, looked much more like a weakness to his contemporaries, rather than the tower of strength that it really was.

'The Theory of the Earth;..laws observable..upon the Earth'

The paper delivered in 1785 was actually published in 1788 with a changed title emphasizing the composition, dissolution and restoration of the land. It was, as might be expected, an expanded discussion of the topics in the abstract, and to geomorphology the interest is limited to the general philosophy of the processes of surficial decay. One interesting feature of the full paper which is completely missing from the abstract, is the flirtation that Hutton reveals with the analogy of a machine for the workings of the earth. That it is a flirtation is evident from the fact that after the first few pages the word 'machine' never reappears, nor is it used in the book of 1795, other than where the 1788 paper is essentially reprinted. Hutton is known to have examined Watt's steam engines in Birmingham in 1774, and at the lead mines at Wanlockhead in 1779, and Playfair assures us than 'he had an uncommon facility in comprehending the nature of mechanical contrivances,..for one who was not a practical engineer.' As it is thought that the 'Theory of the Earth' was completed by 1780, it is natural to expect the machine analogy to enter the manuscript. However, it is clear that the analogy did not rest well with the deeper-seated divine design leitmotif, and Hutton rejected it:

> But is this world to be considered thus as a mere machine to last no longer than its parts retain their present position? Or may it not be also considered as an organised body? such as has a constitution in which the necessary decay of the machine is naturally repaired, in the exertion of those productive powers by which it has been formed.

Hutton realized that the problem lay in powering the machine, and in the fact that Watt no doubt made clear, that parts wear out! Since the system was already driven by divine design, why add another and mundane level to it, for which a more clearly defined power source was needed? That this analogy is missing from the abstract probably indicates that it was written close to 1785, and that Hutton had had second thoughts on the machine analogy.

One item of note, which illustrates the power of Hutton's imagination, is his comment about the Mediterranean:

> Let us but suppose a rock placed across the gut of Gibraltar, (a case nowise unnatural), and the bottom of the Mediterranean would be certainly filled with salt, because the evaporation from the surface of that sea exceeds the measure of its supply (see also I-75).

Deep sea cores have recently revealed that that is precisely what did happen in the late Tertiary (Hsu 1971), and it is a tribute to Hutton as a geologist that he imagined the result two centuries before it was discovered.

In the concluding section of the paper Hutton carefully delineates the true character of what his successors will call 'uniformitarianism.' He attempts to strike a balance between extreme positions as the following extracts indicate:

> We have been reporting the system of the earth as proceeding with a certain regularity, which is not perhaps in nature, but which is necessary for our clear conception... We are not to limit nature with the uniformity of an equable progression.
> We are not to suppose that there is any violent exertion of power, such as is required in order to produce a great event in a little time; in nature we have no deficiency of time.

Later he speaks of 'irregularities' which 'may appear without the least infringement on the general system' and in a different context notes that 'a very small operation of an earthquake' would efface all efforts at recording rates of erosion. From this it is clear that he is trying to strike a balance between gratuitously catastrophic earthquakes or floods, so beloved of most of his contemporaries, and the episodic character of ordinary erosion.

Hutton was surprisingly unsuccessful at finding any information on erosion rates as a basis for estimating 'the age of the present earth,' or, in modern parlance, the length of a cycle. He ransacks the Classical literature for evidence of change along Mediterranean coastlines but can only cite the disappearance of a single small island in the harbour of New Carthage in Spain, mentioned by Polybius. In general then 'we are disappointed..their descriptions.., either give no measure of a decrease, or are not accurate enough for such a purpose.' He compares the difficulty to that of measuring the distance to the stars without the use of a parallax.

He concludes, therefore, that the wasting of the continents is 'extremely slow' and that, in consequence, the 'production of our present continents must have required a time which is indefinite.' The paper comes to a chilling conclusion with the famous phrase which had such an inflammatory effect on his contemporaries:

the result, therefore, of this physical enquiry is, that we find no vestige of a beginning, no prospect of an end.

'The Theory of the Earth, with Proofs and Illustrations'

Critical response to the 1788 paper, primarily on the issues of igneous rocks and the consolidation of strata, persuaded Hutton to prepare the two volume 'Theory of the Earth' published in 1795. Two further manuscript volumes were lost for a century, and some illustrative figures were finally published in 1978! However, these delayed volumes contained little of geomorphological interest. In this section and subsequently, references to the 1795 work indicate the volume and page numbers in the facsimile reprint.

The response to Hutton's geomorphology was limited to vociferous denials, particularly by de Luc, Kirwan, Williams and Richardson, that surface erosion had any long term effect. They admitted the decay of mountains to produce soil, but emphatically rejected any notion that these materials made their way to the sea. Hutton's 1788 paper is reprinted as Chapter I in the book, together with minor alterations and footnotes added in response to criticism. In one footnote, de Luc (1790) is quoted as saying (I-14) that 'you ought to have proved that both gravel and sand are carried from our continents to the sea; which, on the contrary I shall prove not to be the case.' He then proves his case by denying that there is sufficient declivity of the surface, or force in the running water, to move the sediments. Hutton's rather unsatisfactory response, is merely that where there is sufficient slope, or force in the water, then surely de Luc will agree that the sediments 'are travelled upon the land, and are thus carried into the sea - at last' (I-14).

De Luc's attitudes were firmly scriptural, and his biblical blinkers revealed the denudation dilemma, whose resolution was only possible through a denial of erosion. Hutton perhaps felt that the issue was scarcely worthy of detailed debate: his Edinburgh environment was one of sound science in both theory and practice. In Volume II (p240) he touches upon the theme again saying that the issues of whether materials move or rocks are formed (i.e. created) on the surface 'are matters of fact that it is in the power of men who have proper observation to determine; my business is to generalise those facts and observations.' The acceptance of denudation was general amongst those uncommitted to fanciful theories, such as Hutton's friend Joseph Black, but an acceptance of denudation did not mean subscribing to the rest of Hutton's draconian scheme, and Sir James Hall, another close friend, never did accept the full consequences of denudation and erosion, as I shall show later. It is still possible to assert that the major features of the earth's topography derive from more ·spectacular processes in times past: hence the reality of denudation does not mean it has always acted uniformly in the past, and as an additional corollary it does not follow necessarily that geological time extends indefinitely into the past. The concept of eternal time was anathema to Christian theology and de Luc carelessly equated Hutton's indefinite time with eternity. Hutton pointed out the logical error of this supposition (I-221), but it would bring little comfort to his opponents!

A final reason for the muted response to the geomorphological aspects of Hutton's paper is perhaps that, apart from the philosophical issues, he had said very little that was specific. If Hutton had never published his 1795 book, there would have been little record of his thoughts on the minutia of landforms and their origins. As it is, Volume II provides a surprising wealth of detail culled from personal experience and volumes of voyages and travels. One of these books, Saussure's 'Voyages dans les Alpes,' Hutton mined unmercifully: almost half of volume II is devoted to extracts in French from Saussure. The first two volumes published in 1779 and 1786 were available to Hutton: the last two volumes (1796) reached him on his death bed.

To appreciate Hutton's understanding of landforms it is convenient to follow him round the operation of a cycle, just as we might do today in an introductory course. I shall cite selected passages and phrases: his prolixity was such that there are usually several alternate renderings of the same idea in different parts of the book.

Hutton is unspecific on the actual nature of the uplift of strata which initiates the cycle, but he notes that reconstructing the broken strata observed 'with the view of a mineralist, as having all the soils and travelled materials removed' shows that the quantity of erosion that has taken place is 'very great.' The initial uplift of the mountains produces 'ridges of those mountains (which) must have been a directing cause to the rivers' but he is quick to emphasize that rivers 'move bodies with the force of their rolling waters, and wear away the solid strata of the earth' and so 'we must consider rivers as also forming mountains, at least as forming the valleys which are co-relative in what is termed mountain' (II-370). This process is specifically cited for the case of Virginia, and a lengthy quotation from Jefferson describes the passage of the 'Potowmac' through the

Blue ridge. Jefferson ascribes the water gap to the overflowing of ponded waters, while an appendix by his friend Thomson calls upon violent convulsions to achieve the same end. Hutton, however, 'having often admired, in the map, that wonderful regularity,' discerned that 'the original ridges of mountains, or indurated and elevated land, have directed the courses of those rivers, and the running of those rivers have modified the mountains from which their origin is taken.' The case clearly corresponds to the notion of consequent and subsequent rivers, and to the implicit process of superimposition, in terminology which would take another century to emerge.

The ordinary processes of weathering are largely taken for granted, although he does mention, in connection with Alpine summits, that blocks 'are detached by the swelling water upon freezing' (II-305). The chemist in Hutton causes him to describe the breakdown of feldspar in granites to give 'white siliceous earth' or kaolin and he willingly submits his theory to the proof that wherever kaolin is found 'there should also be granite or feldspar to explain its origin' (II-251). This process causes 'a species of sand' to remain (II-305). In general, 'calcareous, argillaceous, and other soluble earths compose many of the strata ; but in many more, which are partly or chiefly composed of insoluble substances these soluble earths are mixed in various proportions' (II-94). Insoluble materials are left as ore residues: stream tin in Cornwall, stream gold in Peru (II-253).

The decay of rocks to give soil is viewed as a continuous process because, contrary to views of de Luc, fertility must be renewed with new material. Hutton the farmer was well aware of this and well knew that 'over and over a thousand times may be repeated this alternate possession of the transferable soil, by moving water on the one part and by fixed vegetation on the other, but at last all must land upon the shore, whether (sic) the river tends' (II-209). In addition to 'seeing the powers which are employed in thus changing the surface of the earth, we must also observe how their force is naturally augmented with the declivity of the ground on which they operate' (II-565).

Hutton, even in this full exposition of his 'Theory,' seemed unable to come to any useful estimate about the 'age of the present earth' from estimates of erosion. One specific example is quoted from the Pyrenees where Gensanne estimated a lowering of ten inches a century and and a predicted lifetime for the Pyrenees of a million years. The French source in which Gensanne's estimate is given is sceptical, and Hutton remarks 'I do not know in what manner M.Gensanne made his calculation; I would suspect it was partial, and not from general observation' (II-145). Likewise he draws no conclusions from an example on the Normandy coast where he cites an authority for the retreat of the cliffs at one foot per year (II-263).

His conclusion on this matter is, therefore, no improvement over his earlier work, although he displays some signs of hope:

> - some data may perhaps be found by naturalists and antiquarians, when their researches shall be turned to this subject. It is only in this manner that there is any reasonable prospect of forming some sort of calculation concerning that elapsed time in which this present earth was formed, a thing which from our present data we have considered as indefinite.
>
> In this view .. nothing is more interesting than the beds of rivers..(II-99).

In the circumstances, it is surprising that he does not include travellers' accounts of Niagara Falls. The retreat of the waterfall along the gorge was frequently used as a natural chronometer of the age of the earth, but the first such estimates (see p97) appeared about the time that Hutton was writing his book. A brief mention of the Falls is made by Playfair (1802), writing a few years later (p97, 98).

Irrespective of the rate of surface erosion, Hutton was in no doubt that the overall effect was a lowering of the land surface and he gives a clear description of terraces, or 'haughs.' 'It is not infrequent to see relicts of three or four different haughs which had occupied the same spot of ground upon different levels, consequently which had been formed and destroyed at different periods of time' (II-210). He goes on to cite examples from the Ganges and Babylon.

That the course of fluvial activity will not always run smoothly he illustrates by noting that 'it may not be infrequent, perhaps, to find a small stream of water in places where a greater stream had formerly run; this will naturally happen upon many occasions, as well as the opposite, by the changes which are produced upon the form of the surface' (II-497). In cases like this, one wishes that Hutton had cited actual field sites since such a quotation could be 'over-interpreted' as an insight into misfit streams or a hunch about formerly glaciated valleys. In reality it may relate to something more prosaic, perhaps an observation of shifting streams and tributaries on a floodplain.

All surface operations combine to 'form a system of rivers and their branchings..increasing in their size as they are diminished in their number by the uniting of their waters' (II-446). This system 'must gradually produce valleys, by carrying away stones and earthy matter in their floods.' He observes:

> that no such system could arise from the operations of the sea when covering the nascent land; (and) that

however this sytem shall be interrupted and occasionally destroyed, it would necessarily be again formed in time, while the earth continued above the level of the sea. Whatever changes take place from the operation of internal causes, the habitæble earth, in general, is always preserved with the vigour of youth, and the perfection of the most mature age (II-538).

Hidden away in this quotation is the coy reference to 'internal causes' which is as far as Hutton dare move towards a recognition that endogenetic processes, manifest as earthquakes, may have a part to play. The Lisbon earthquake in 1755 caused lakes to oscillate in Scotland and Sweden, and was widely reported so that he would be well aware of their power. In a footnote to an account of earthquakes around Comrie, Perthshire (Taylor 1794), Hutton is cited as remarking that the area is 'where the primary strata sink below the surface, and are covered by the secondary, or horizontal strata,' - in modern terms, the Highland boundary fault. In addition, the Classical literature, which he had scanned for evidence of erosion, was replete with descriptions of their effects. Regrettably, earthquakes were the first resort of every fanciful theorist in search of instant topography and Hutton could not afford to be seen in that camp, or all the theoretical power of gradual fluvialism would evaporate.

In time, all 'travelled materials' would arrive at the coast and 'countries are then formed at the mouths of rivers in the sea, so long as the quantities of materials transported from the land exceeds that which is carried from the shore, by tides and currents, into deeper water' (II-96). Coastal erosion on a large scale is illustrated on the Caithness coast for 'along this coast there are many small islands, pillars and peninsulas of the same strata, corresponding exactly with that which forms the greater mass.' Looking at the coast of north-west Europe in general, 'as well as the most particular maps,' Hutton suggests that outlying islands hint at the 'former extent of our continent and land around,' and he connects Ireland with Britain, Orkney with Scotland, England with France along the chalk coastlines, and finally Shetland with Norway, although this latter 'is not proposed as a fact immediately supported by natural appearances; it is only to be considered as an enlarged view in which we may contemplate the operations of this earth upon a more extended scale; one which may be conceived as a step in our cosmogeny' (II-285).

Relative changes of the land and sea are revealed because 'from the south to the north of this island there are, in many places, the most evident marks of the sea having been upon a higher level of the land; this height seems to me to be about 40 or 50 feet perpendicular at least, which the land must have been raised' (II-165). Evidence is then cited from the Thames valley, Suffolk, the 'Frith of Forth,' Kinneel, east of Edinburgh, Cromarty and Fife; all in association with 'sea shells.'

Even with the use of the 'Theory' 'it must not be imagined that, from the present state of things, we may be always able to explain every particular appearance of this kind which appears; for example, why upon an eminence, or the summit of a ridge of land which declines on every side, an enormous mass of travelled soil appears; or why in other places, where the immediate cause is equally unseen, the solid strata should be exposed almost naked to our view' (II-98). In this example the 'Theory' runs up against the problem of erratics; one manifestation of glaciation. I shall return to this after a consideration of the problem of lakes.

Within the framework of the 'Theory' Hutton knows that 'lakes are not in the natural constitution of the earth..they must be formed by some posterior operation' and for the embarrassingly large Lake Geneva 'we are therefore led to believe, that the passage of the Rhone through the lake, in its present state, is not a thing of long existence, compared with the depredations which time had made by that river upon the earth above the lake' (II-147), although he leaves the actual cause to 'those who have had the proper opportunity of examining that country.'

Hutton's intuition was correct and three causes are suggested for lake basins: (i) blockage by landslides (ii) 'the operation of an earthquake, which may either sink a higher ground or raise a lower, and thus produce a lake where none had been before' and (iii) 'the dissolution of saline or soluble earth substance which had filled the place' (II-149). This is one of the few times that Hutton calls upon earthquakes by name, and it perhaps indicates the seriousness with which he viewed the problem of lakes; he quotes Pallas on the lowering in the level of the Caspian but has no explanation for the cause.

The problem of lakes remained a weak spot, and the Swiss born de Luc in particular thought that the lack of infill in Lake Geneva argued for the recent origin of all Alpine topography. He wondered why, if Hutton thought the lake could be created by an earthquake, the argument would not hold for all the alpine valleys?

The resolution of this problem lay with glaciation, and even though Hutton knew of present glaciation, through the writings of Saussure (1779, 1786), he was not able to extend this to the excavation of lake basins, even in his powerful imagination. However, he did apply the idea the distribution of erratics, and particularly the case of extremely large granite blocks. I have shown recently (Tinkler 1983b) that a good case can be made for Hutton not only applying the idea of glaciation to the transport of granite blocks from the Mt

Blanc massif to the Saleve and the Jura, but also to the dispersion of granite blocks in Britain. Hutton's case is almost as dispersed as his erratics, but it is central to the argument to realize that the 'Theory' required the uplift of strata from the sea bed to form the initial mountain topography. Hutton has already described this, and deduced that enormous erosion has taken place since that time. By reversing the process in the imagination, the mountains were formerly higher and colder (Hutton in other work reported by Playfair (1805) had estimated the lapse rate of temperature with altitude quite accurately) and it follows that they were covered originally with more extensive glaciers:

> There would then have been immense valleys of ice sliding down in all directions towards the lower country, and carrying large blocks of granite to a great distance, where they would be variously deposited, and many of them remain an object of admiration to after ages, conjecturing from whence, or how they came. M. de Saussure, who has examined them carefully, gives demonstration of the long time during which they have remained in their present place. The lime-stone bottom around being dissolved by rain, while that which serves as the basis of those masses stands high above the rest of the rock, in having been protected from the rain. But no natural operation of the globe can explain the transportation of those bodies, except the changed state of things arising from the degradation of the mountains (II-218).

Hutton had previously remonstrated with both Saussure, who believed in the agency of a huge debacle or flood, and de Luc who invoked violent volcanic explosions, over their explanations of the erratics. As he pointed out 'if these dispersed blocks of stone are to be explained by explosions, there must have been similar explosions in other countries where there is not the smallest appearance of volcanic eruptions, for, around all our granite mountains, and I believe all others, there are found many blocks of granite, travelled at a great distance, and in all directions' (I-403). Although there is no mention of ice in this last passage the connecting link can be made with the opening sentences of Chapter VII (II-212) where the reader is enjoined, in the context of great granite erratics, 'to enlarge our thoughts with regard to things past by attending to what we see at present, and we shall understand many things which to a more contracted view appear to be in nature insulated or without proper cause.'

Hutton's perceptive appreciation of the power of glaciation is a fitting conclusion to this review of his geomorphological thought for with it he foreshadows the very process that will be accepted, sixty years later, as finally removing the remaining objections to fluvialism. Hutton was in need of nothing so much as a first class editor, and it is fortunate for science and literature that already in Edinburgh one was waiting in the wings: the Reverend John Playfair.

JOHN PLAYFAIR (1748-1819)

James Hutton died in 1797 with a manuscript on agriculture still unpublished. Davies (1969) has pointed out that, embedded as he was in the eighteenth century, his motivating concern was to resolve the denudation dilemma. This he did by the novel method of permitting denudation, but discovering a restorative process that enabled a habitable globe to persist indefinitely. Certainly, a concern with divine design is threaded through his writings, but perhaps it is fortunate that the 'Theory' Hutton developed so persuasively from the field evidence so readily permitted the successful separation of theology and geology, and left him filled with joy at the sublime benevolence so revealed.

John Playfair, in contrast, twenty-two years younger than Hutton, never allowed the two concerns to mix. He was born in Benvie, Forfarshire, in 1748 and was educated by his father, a minister. His talent was for mathematics although he trained as a minister at the University of St Andrews. He left the University in 1769 and lived in Edinburgh until 1773 when he obtained his father's living at Benvie. It is asserted by his biographer, Jeffrey (in Playfair 1822), that he became acquainted with Hutton, Black and Adam Smith during this stay in Edinburgh but Playfair himself (1802), in a footnote to Hutton's biographical memoir, notes that he first became acquainted with Dr Hutton about 1781. From 1773 to 1782 he occupied his father's living at Benvie, although a geological interest may have surfaced in 1774 when he met the Astronomer Royal who was studying the effect of the mountain Schehallien on a plumb line. Much later Playfair (1811) made a detailed lithological survey of Schiehallien.

From this point onwards there is little record of an active interest in geology until the death of Hutton in 1797. However, Playfair was a part of the active, but informal, group that centred around James Hutton, Joseph Black, Adam Smith and from about 1788 onwards also included Sir James Hall, who was already proposing to test Hutton's 'Theory' by controlled experiments melting and re-crystallising whinstone (basalt).

Almost immediately on Hutton's death, Playfair began to 'draw up an abstract' of his works 'with a view to the composition of a biographical memoir; an occupation which eventually gave birth to the 'Illustrations of the Huttonian

Theory of the Earth' (Jeffrey, in Playfair 1822).

It was probably at this time that Playfair realized how confused and repetitive the two volumes of the 'Theory' really were. Playfair had to re-organize the whole, merely to understand it, and it is significant that the 'Illustrations' appeared in 1802, whereas the biographical memoir was read to the Royal Society of Edinburgh in 1803, and was published in 1805.

'Illustrations of the Huttonian Theory of the Earth' (1802)

I provided a detailed summary of Hutton's writings in an earlier section because, in the rush to praise the lucidity of Playfair, it has often been forgotten that most of his passages are paraphrases of corresponding passages in Hutton's 'Theory.' Playfair never pretended otherwise, his intention was 'little more than a clear exposition of the facts, and a plain deduction of the conclusions grounded on them; nor shall I claim any merit to myself, if, in the order which I have found it necessary to adopt, some arguments may have taken a new form, and some additions may have been made to a system naturally rich in the number and variety of its illustrations.'

No complete and detailed comparison appears to have been made between the 'Theory' and the 'Illustrations,' and there is no space for one here. The 'Illustrations' were recognized for their literary style immediately. Humphry Davy, in his Royal Institution lectures of 1805 (Siegfried and Dott 1980), said that 'whatever may be the fate of the opinions of Dr Hutton, the work of the distinguished Professor of Natural Philosophy of Edinburgh considered as a literary production will stand the ordeal of time and of the most rigid criticism.' Here I shall merely illustrate the difference in treatment, and note a few of the instances where Playfair begins to develop or defend the theory. The # numbers in this section refer to the numbered paragraphs in Playfair (1802). The reader is warned that, due to a proof reader's error, #134 is repeated in that work, but it was corrected in the 1822 reprint contained in the 'Collected Works.'

A nice example of Playfair's style in contrast to Hutton's is provided by the example of the 'haughs':

#100 The changes which have taken place in the courses of rivers, are also to be traced, in many instances, by successive platforms of flat alluvial land, rising one above the another, and marking the different levels on which the river has run at different periods of time. Of these, the number to be distinguished, in some instances, is not less than four, or even five.

Similarly, the so-called Playfair's law of accordant junctions is largely culled from a maze of Huttonian statements expressing the same idea:

> #99 Every river appears to consist of a main trunk, fed from a variety of branches, each running in a valley proportioned to its size, and all of them together forming a system of vallies, communicating with one another, and having such a nice adjustment of their declivities, that none of them join the principal valley, either on too high or too low a level; a circumstance which would be infinitely improbable, if each of these vallies were not the work of the stream that flows in it.

Playfair supplies the observation that the slopes meet smoothly at the junctions, and the mathematician in him provides the final punch line, but the rest is from Hutton.

Playfair adds quantitative data to the arguments wherever possible, and so he cites an authority to show that water in a flood 'contains earthy matter suspended in it, amounting to more than the two hundred and fiftieth part of its own bulk' (#103). He reminds readers in the Notes (#349) that fragments of rocks 'are rendered specifically lighter by the fluid in which they are immersed, and lose by that means at least a third part of their weight: they are at the same time, impelled by a force proportional to the square of the velocity with which the water rushes against them, and proportional to the quantity of gravel and stones which it has already put in motion.' There is abundant evidence here that Playfair was applying his mathematical and physical skills to a problem in hydrodynamics, and his library contained all the eighteenth-century classics in these fields: Euler, the Bernoulli family productions, and the French school of Lagrange, Laplace and Legendre.

The fluvial idea does have difficulties, and Playfair, in a lengthy footnote (to #361) which is usually ignored, draws attention to the problem of the Great Glen from Fort William to Inverness. After a detailed description he notes its sympathy with 'the vertical strata which compose the mountains on either side,' and is definite that 'here, then, we have a valley, not cut out by the working of any streams which now appear.' He proceeds to test the 'debacle' hypothesis as an explanation of its origin. He points out that a column of water acting against a cross-section of the valley would, to excavate it, have to 'overcome the cohesion and inertia of a column of rock of the same section and of the length of sixty-two miles.' The objection is devastating. He then proffers the alternative idea that at least its bottom was scoured by tidal currents at the relatively recent time, documented elsewhere in the book, when the land lay one hundred feet lower with respect to the sea. The solution is ingenious but hardly convincing.

Playfair adds considerably to the section on apparent changes in the level of the sea and produces evidence from the Baltic of the actual rate of change as measured from markers by Celsius: 40 inches per century. In addition, he shows that physically it is easier to displace a relatively small land mass through 30 feet, than to do the same for the entire ocean, despite the difference in relative density. This is supported by evidence from the Mediterranean showing that there the land has sunk relative to sea level in recent times, whereas it has risen in northern Europe.

On the vexed question of Lake Geneva he is no more successful than Hutton, indeed it is perhaps the one occasion where Hutton is more lucid. However, he compensates for this by providing a well organized section on the 'Transportation of Stones..' in which Hutton's glacial arguments are clearly stated vis-a-vis the alternatives of torrents, rivers and debacles and, as I have shown elsewhere (Tinkler 1983b), Playfair names the location of a glacial erratic in Brodick Bay, Arran, transported from the summit of Goatfell; a convincing first application of the glacial idea to Britain.

Continental trip and the projected second edition

The previous section can only hint at the overall excellence of the 'Illustrations,' in both style and content. Devoid of Hutton's design leitmotif, except in direct quotations from Hutton, it speaks directly to the modern reader. When he had completed the 'Illustrations,' it is easy to imagine that Playfair was burning to visit the continent to examine some of the puzzling field sites: Lake Geneva, the Jura and the Saleve erratics, the glaciers so clearly described by Saussure and the coastal sites in Italy. He intended to make the trip in 1803, but before he could, war broke out again and he had to wait until 1815-6 to make the excursion.

One object of the proposed trip was to gather material for a second edition of the 'Illustrations.' All we know of this work is the projected plan to describe the 'authenticated facts' of geology, and from these to draw general inferences to be tested against various geological systems. Those found wanting on any item were to be rejected. The system remaining consistent with the facts would then be applied to the phenomena of geology (Jeffrey 1822).

The approach is not foolproof, but it would doubtless have been successful in Playfair's sure hands. When his continental trip was eventually made it lasted eighteen months and a reasonably full account of it is given by Jeffrey (1822). The item of primary interest is that Playfair realized some of his errors in applying the glacial idea. Whereas previously he had required a continuous land slope from Mt Blanc to the Jura, he soon realized that:

60

A glacier, which fills up valleys in its course, and which conveys the rocks on its surface free from attrition, is the only agent we now see capable of transporting them to such a distance, without destroying the sharpness of the angles so distinctive of these masses. That mountains formerly existed of magnitude sufficient to give origin to such extensive glaciers, is countenanced by other phenomena observed in the Alps, and does not imply any alteration in the surface, so great as the supposition of a continued declivity between the two extreme points, which is, after all, insufficient to remove the objection arising from the sharp angles of these rocks.

While he does not describe the 'other phenomena' it is easy to infer that these probably included the series of terminal moraines marking the outer limits of Pleistocene glaciation. This observation would have only antiquarian interest if it were not that it is contained in Jeffrey's account of his life, together with a reprint of his 1802 'Illustrations,' (Playfair 1822). At this time, geology in Britain was extremely active. Charles Lyell was making his mark in London scientific circles, and was about to embark on the writing of his 'Principles of Geology.' It is known from the error in the paragraph numbering, corrected in the reprint, that Lyell was using the reprint. It can be no coincidence that Lyell has a chapter entitled 'Prejudices which have retarded the progress of Geology' while the very last Note in Playfair (#445 to #459) is entitled 'Prejudices relating to the Theory of the Earth.' Lyell, seeking inspiration for the structure of his book, may well have looked at the recently reprinted version of the classic in the field. The plan of the projected second edition may well have suggested to him that a comprehensive review of earlier theories would also set the scene for his own revelations.

By 1822 certainly it was true, that for geology as a whole, even the elegant 'Illustrations' were dated, at least with respect to the multitude of new field evidence and William Smith's geological map of England and Wales (1815). But, in the narrower field of geomorphology the battle had hardly begun. As Davies (1969) remarks, 'some portions of the 'Illustrations' might still be read with profit by those seeking an introduction to geomorphology' but the fact that 'in its day the book was unrivalled as a geomorphic text' was lost on Playfair's contemporaries.

Contemporary reactions to the 'Illustrations'

Davies (1969) has argued that geologists, aware of the fight that the Plutonic view of geology engendered, imagined that Huttonian geomorphology was also under fire, whereas 'in the geomorphic context the book had no inflammatory effect whatever.' In fact, a compromise position would be more accurate. Murray (1802), in his immediate, and Wernerian, response to the 'Illustrations,' grudgingly accepts denudation, and that sediment is transported offshore. 'What (time) must be required to level these mountains with the sea?' he asks, 'Millions of years would not suffice.' Murray denied, though, that there was any subsequent process of restoration. Humphry Davy, who met Playfair in 1804, wrote in his Royal Insitution lectures for 1805, and presumably in later lectures to the same body, that:

> it is impossible to doubt that in the lapse of ages some slight diminution of our mountains must take place from the agency of rivers, but there are many causes in nature which appear to counteract the effect. Forty centuries nearly have passed away and no material change in the form of the solid matter of the globe has taken place in consequence of this agency. From the most accurate observation it appears nearly a hundred years would be consumed before a single inch of a rock like that which this instance represents, (Davy showed a specimen at this point) or of Mount Blanc, would be lost. And with regard to the ultimate termination of such effects, it is almost ridiculous to reason seriously (Siegried and Dott 1980).

This particular extract may have been aimed at Buffon's 'Epoque de la Nature' (1779) where the age of the globe is estimated at 93,000 years, rather than at Playfair, but the point is just as apt.

Clerical opinion, usually so vehement, could be favourable. Hugh Miller, the Scottish geologist of 'Old Red Sandstone' fame, quoted from a sermon by the Reverend Thomas Chalmers, given in 1804, where he remarked that:

> there is a prejudice against the speculations of the geologist, which I am anxious to remove. It has been said that they nuture infidel propensities. It has been alleged that geology, by referring the origin of the globe to a higher antiquity than that assigned to it by the writings of Moses, undermines our faith in the inspiration of the Bible, and in all the animating prospects of immortality which it unfolds. This is a false alarm. The writings of Moses do not fix the antiquity of the globe (Miller 1854, p115)).

At about the same time, and possibly before, Sir James Hall, the experimental geologist, expressed his belief in the efficacy of 'vast torrents flowing with great rapidity' as necessary for the explanation of some aspects of surface features, and in this view he allied himself with Pallas, Saussure, and Dolomieu (Hall 1805). The first published hint of this is given in Playfair's 'Illustrations' where he defends, in #364, the idea that the landscape around Edinburgh has been shaped by a torrent. He does not name the source of the idea, but it is exactly the area treated in Hall's subsequent paper (Hall 1815), which I discuss in Chapter Six. In #367 Playfair refers to Hall's forthcoming paper (Hall 1805) in which a footnote announces this view (see Chapter 1, p9).

The 'Theory' was about to be put to tests to which it might not have an answer, at least as formulated. Nevertheless, it must be said that it was raising some controversy, however local and sporadic.

More evidence of local reaction is provided in a paper by Allan (1815). During a geological investigation of the Faroe Islands, he was led to consider the deeply eroded dykes:

> They are considered by LANDT as marks of some violent convulsion in nature; and to a common observer, they have very much that appearance. The effects of the weather, however, without any such assistance, are quite sufficient to accomplish this end; nor will its operations be tardy; the constant action of the surface-water on the summit, and the continued lashing of the waves at the base, are agents of sufficient power; and we have thus dykes washed from their sockets, for an extent of several hundred feet, leaving a frightful chasm in rocks of enormous height.

Here one senses Nature illustrating Huttonian doctrines for Allan (who had had bad weather for a month!).

In London, as well as Edinburgh, the new ideas were not neglected, not least because de Luc was still writing geology. An anonymous reviewer, in the 'Annual Register' for 1809, noted that de Luc's 'Travels in Northern Europe' was a continual 'controversy with Mr Playfair.' Hutton and Playfair are numbered amongst those geologists who 'ascribe valleys and vales to the mechanical action of running water, and the indentations of the coasts to that of the sea.' De Luc apparently considered this to be 'the greatest (controversy) which has arisen amongst geologists; a controversy very extraordinary, since, after the length of time employed in observing the earth, it might seem that this question should have been decided by facts the most common and the most simple.' That, of course, was exactly Hutton's point quoted earlier: the difference between them was ideological.

Robert Bakewell, in his 'Introduction to Geology' (1813), included the fluvial idea in his review of the causes of valleys and concluded that 'geologists seem now generally agreed, that the action of rivers is not sufficient to explain all the phenomena of valleys.' Likewise Greenhough (1819) quoted Playfair at length but rejected a fluvial origin for valleys.

A final barometer of the progress of fluvialism, and the 'Theory' in general, is provided by the 'Summary of Physical Geography' which prefaces Thomson's 'New General Atlas' (1817). This was probably written by Thomson himself (Margaret Wilkes pers. comm. 1982) and steers a middle course to such an extent that the origin of springs is still attributed, in part, to the 'ascent of subterranean vapours,' and one of the causes of islands is 'some violent convulsion, probably that of an earthquake, which has torn off portions from the adjacent continent.' Yet, in the next breath it is admitted that 'it is possible, that, in the appearance and constitution of islands we behold the rudiments of new continents, destined in future ages to supply the place of those which now exist, but which have already commenced, by a slow but sure process of decay, to return to the depths of the oceans': a truly Huttonian perspective.

A compromise position is once more revealed; the reality of denudation, once accepted, begins to force the moderate mind to an acceptance of its long term consequences, unless of course one has preconceived notions:

> His celebrated Theory of the Earth might have been more worthy of notice had he not seemed too much to overlook the agency and providence of the Deity, by referring entirely to the operation of second causes, the continual tendency to decay and the sources of renovation that he imagined he discovered in the mundane system (Reverend A. McConechy 1845).

It is rather unfair to leave Hutton and Playfair with a body blow from a cleric, especially one based in Hutton's own county, but it has to be admitted that it was not untypical of attitudes that persisted well into the nineteenth century.

The eradication of such dogmatism needed the new forces of professionalism and reasoned discourse; precisely those Baconian qualities that Playfair extolled in Hutton. But, a theory stands on its own feet:

> If, therefore, the science of the present times is destined to survive the physical revolutions of the globe, the HUTTONIAN THEORY may be confirmed by historical record; and the author of it will be remembered among the illustrious few, whose systems have been verified by the observations of succeeding ages, supported by facts unknown to themselves, and established by the decisions of a tribunal, slow, but infallible, in distinguishing between truth and falsehood (Playfair #134 1802).

Part III

THE PRINCIPLES: A CENTURY OF DEBATE

An attempt to explain the former changes of the earth's surface by reference to causes now in operation.

(C. Lyell, 1830)

Changes of the surface at Fra Ramondo, near Soriano, in Calabria.

1. Portion of a hill covered with olives thrown down.
2. New bed of the river Caridi.
3. Town of Soriano.

Chapter Four

THE BACKGROUND OF INDUSTRY, SCIENCE AND SOCIETY

> *Have I ever told you of the wonderful & surprising curiositys we find in our Navigation?*
>
> (Josiah Wedgewood, 1767)

> *It is to the mining business chiefly that we are indebted for that demonstration of which I shall now give an example.*
>
> (James Hutton, 1795, II-289)

810 -1420

The early years of the nineteenth century saw radical changes in the approach to geology; Zittel (1901) labelled the years 1790 to 1820 the 'Heroic Age of Geology.' Rarely do such changes take place entirely divorced from the rest of their scientific, economic and cultural milieu, and this brief chapter outlines some of these background factors, and their influence on geomorphological thought, from the mid-eighteenth to the late nineteenth century, without in any way trying to explain their origins.

THE INDUSTRIAL REVOLUTION

The theoretical ideas of most late eighteenth-century geological writers, Hutton and Playfair excluded, were hardly changed from those of the century before. Their writings, however, did reveal one great difference, and that was a detailed acquaintance with geology in the field. The difference was no accident; it was just one consequence of the Industrial Revolution. Williams (1789), an opponent of Hutton's views, was both a lead and coal mine manager, and his book, for all its theoretical defects, was very sound on the stratigraphy of coal measures. Whitehurst, horologist and sometime architect, was quite Williams' equal in absurd theoretical speculations (Whitehurst 1778), but he had a clear view of Derbyshire stratigraphy. As a member of the informal Lunar Society (Schofield 1963, Ritchie-Calder 1982), known for its impact on Midland industry, he surprised Josiah Wedgewood, the potter, with his extreme theological views when his work was eventually printed.

Canals and mining

The roots of the Industrial Revolution can be traced, in part, to the development of the steam engine from the initial, and extremely inefficient Savery engine (1698) through the Newcomen self-acting model (1712), to the greatly improved Watt version (1765) whose commercial model (1774) stimulated Hutton's brief flirtation with the machine analogy as a model for the 'Theory of the Earth.' The power of steam enabled the mechanization of industry, and the demand for raw materials fostered the growth of Britain's canal network from 1759 to the 1790's. The mere mechanics of expanding mining to new scales of production, and of building canals across the extremely varied topography of England, inevitably focussed attention on the near surface geology. Large landowners, anxious to exploit their actual, or potential mineral wealth financed mineralogical investigations and the building of canals. The result, for geology, was the rise of a class of men, whom we would now classify as consultants, able to give a shrewd evaluation of the local geology; the likelihood of finding coal, or the ease of routing a canal. Josiah Wedgewood, who as a potter had a professional interest in clays and earths, wrote in a letter to Thomas Bentley (1767), of the 'wonderful & surprising curiositys we find in our Navigation' with reference to the newly-begun Mersey to Trent canal. John Williams (c1730-1797), William Smith (1769-1839) and Robert Bakewell (1768-1843) were just such men, and James Watt and James Hutton were both involved with canal projects in Scotland.

Canals were built much earlier in France, but not for industrial purposes, and with little apparent effect on geologists. In the United States canals came much later but it is noteworthy that Amos Eaton's 'Survey of the Canal Rocks' (1824) for the proposed Erie Canal was the first detailed stratigraphic investigation in New York State, and probably in the United States (Hall 1843).

Mines, whether for lead, iron, or coal, revealed fractured, faulted and folded strata, and these, together with the creeps and bursts so typical of mines, perhaps led to an unfortunate emphasis on the great energies that must have been expended to create them, apparently in one fell swoop. Miners in Derbyshire during the Lisbon earthquake (1755), for example, 'felt the rocks move, and heard noises, which were scarcely perceived by those above' (Bakewell 1838). It can hardly be doubted that such experiences would tend to promote a catastrophic, rather than a gradualist, perspective on geological processes. In this respect, therefore, the new field knowledge had an unfortunate effect on geomorphology. But at the same time it revealed another unsuspected feature of surface form. Hutton's elucidation of unconformities demonstrated great revolutions in geology (the laying of new sediments on the upturned edges of folded strata) but it had not been applied conceptually to the present surface. The British coalfields are notorious for their heavily faulted structure, no sign of which is revealed at the present surface. The significance of this fact seems to have revealed itself slowly. Robert Bakewell, the mining consultant and geologist, published his 'Introduction to Geology' in 1813, but it was not until the fifth edition in 1838 that he added a special chapter dealing with 'the general removal and disappearance of the coal strata raised by faults above the surface of the ground.' He remarked that Williams, 'a practical miner,' appeared not to have noticed the removed strata, but quoted Farey (1815), and Mammott, a former colleague, who had. In a rueful footnote, he willingly acknowledges part of the blame for failing to realize, over thirty years previously, that 'it was of such general occurrence, and .. like others, in calling the removal a case of denudation.' In this context it should be noted that 'a denudation' had the specific connotation of strata locally removed, during most of the nineteenth century. Bakewell also wished to object to the agent of removal usually assigned, whether local or general - 'the vague supposition of diluvial currents' - and he replaced it with the submarine removal of the strata before it was raised above sea level.

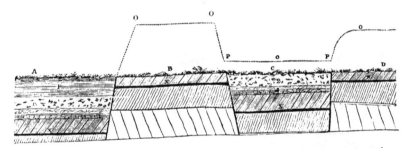

A B C D. The surface of a coal-field, in which the strata are dislocated and elevated by faults.
The dotted line O O P O P O, marks the space that the upraised strata would have filled, had they not been removed by some unknown cause.
X X X X'. The same bed of coal at different depths.

Figure 3. *The figure in Bakewell's "Introduction to Geology" (5th ed. 1838) illustrating the general removal of the coal strata.*

Bakewell was right to call attention to the fact, and it is not one which could be inferred from a superficial inspection. His own experience shows that the generality of the loss which, he reasoned, must apply to other strata too, (otherwise why did the speculative inundations 'select for their theatre of action all the coal-field districts of Scotland, England and Wales?'), was only realized after considerable experience. It was this vast experience, the mere act of seeing a great many exposures in many different places, that was such an enormous catalyst to geology in the early nineteenth century.

William Smith, well known as the 'Father of English Geology' and for his delimitation of the broad stratigraphy of England, was another practical geologist who travelled all over England, expending all his earnings in the process. The result, in 1815, was his magnificent 'Geological Map of England and Wales, with part of Scotland; exhibiting the Collieries, Mines, and Canals, the Marshes and Fenlands originally overflowed by the Sea; and the Varieties of Soil, according to the Variations of the Substrata; ..By William Smith, Mineral Surveyor.' The map was at the scale of five miles to the inch, and required fifteen sheets. His immediate impact on geomorphology was, perhaps, negligible, although it is said that in building sea walls and draining landslip sites, he mimicked natural landforms whenever possible (Geikie 1905). However, like Bakewell, his practical work bore fruits for geology as a whole that could not be ignored. Neither man joined the Geological Society, but both were well known in British geological circles. Their practical experience was much sought after, and their example showed what had to be done to accomplish significant results. Smith based himself in London in 1805 and promoted, unsuccessfully, the idea that he should be attached to the Ordnance Survey; thirty five years before the Geological Survey was instituted, in precisely that manner.

Maps and surveys

By the time the railway era was ready to begin (1830), a sound practical knowledge of British geology was available. The rising demands of industry, and the long period of troubled foreign relations, from the beginning of the French Revolution (1789) until the Treaty of Versailles in 1815, encouraged another field of activity: the construction of accurate maps.

English cartography reached an aesthetic zenith in the eighteenth century with the production of county maps at the scale of one inch to one mile. On the other hand, Sir Joseph Banks, President of the Royal Society, in 1791, compared English County maps unfavourably to Rennell's map of Bengal (Arden-Close 1926). The first report of the Ordnance Survey was able to quote an example, from Taylor's map of Dorsetshire (1761), of a church spire in error by three miles in eighteen, and ten percent errors were not uncommon. The Ordnance Survey, as it came to be known, was officially founded in 1791, after a prior existence in the person of General Roy during which primary baselines were measured. The initial imperative was to connect the Greenwich and Paris Observatories by ground triangulation, but this soon developed into a plan to produce accurate one inch published maps of English Counties, admittedly, in the first instance, for primarily military purposes. However, there was no secrecy and the first map was published of Kent in 1801, followed by Essex in 1805. It was in this light that Playfair wrote, in the concluding section of the main text of the 'Illustrations,' that from the:

> advancement .. of other sciences, less directly connected with the natural history of the earth, much information may be received. The accurate geographical maps and surveys which are now in the making; the soundings; the observation of currents; the barometric measurements, may all combine to ascertain the reality and to fix the quantity of those changes which terrestial bodies continually undergo. Every new improvement in science affects the means of delineating more accurately the face of nature which now exists, and of transmitting, to future ages, an account, which may be compared with the face of nature as it shall then exist (Playfair 1802, #134).

Playfair wrote these lines sensitive to Hutton's frustration at being unable to find useful estimates of the length of the 'present earth' from comparative studies. In addition he had made a lithological survey of Schiehallion which tried to quantify the amount by which a plumb bob is deflected by the gravitational mass of a mountain. On his death, his library included Rennell's maps of India, plans of various proposed canals in England and Scotland, eighteen sheets of the Ordnance Survey, a three foot telescope, and a theodolite.

The idea for a Geological Survey grew slowly out of the Ordnance Survey. Macculloch (1773-1835) got himself appointed as mineralogical and geological surveyor to the trigono-metrical section of the Survey in 1814, with the particular duty of advising on the effect of anomalies due to topography. In 1826 he was instructed by the Treasury to prepare a geological map of Scotland which he did, in the absence of Ordnance maps, by colouring the Arrowsmith map. The map was still unfinished in 1835 when he died in a carriage accident, on his honeymoon in Cornwall. It was published in 1840.

In 1835 a committee of English geologists (Lyell, Sedgwick and Buckland) was appointed to advise on colouring the Ordnance Survey maps geologically. The outcome was the establishment, in that year, of a section attached the Ordnance Survey and under the direction of H.T. de la Beche, who had already been working informally towards the same ends. In time, the Geological Survey became independent of the Ordnance Survey apart from the use of Ordnance base maps at various scales. The Geological Survey naturally focussed on areas of economic interest so that its direct impact on landform studies in the nineteenth century was small, except

71

when officers had a personal interest in landscape. Published memoirs usually provided only a meagre section on superficial deposits and 'drift.' Nevertheless, the basic inventory of geology was a necessary basis for all branches of the subject, and many influential papers on geomorphology came from the pens of survey officers: the Geikie brothers, Archibald and James, A.C.Ramsay and J.B.Jukes.

On the continent, a French national geological map was prepared in the years 1825 to 1840, upon a methodology borrowed from England and based on Greenhough's geological map of England and Wales (1820). Across the Atlantic, William Maclure, an expatriate Scot, published the first geological map of the Eastern United States in 1817, and by 1830 the industrial imperative of oil and coal saw State Geological surveys beginning to be established; indeed by 1840 almost all States east of the Mississippi had their first survey completed or underway. In the same decade, a number of federally financed explorations took place in areas west of the Mississippi. The State Surveys were either maintained or revived in later decades as additional demands were placed upon them. In North America, however, the vast size of the continent slowed down topographical mapping, and when the United States Geological Survey was finally established in 1879, topographical survey was supported only to the extent needed for geological studies (U.S.G.S. 1974).

Field mapping was a slow business, and, for some field problems of a geological nature, scales much larger than those published were needed to comprehend the problem. The Ordnance Survey adjusted slowly but eventually began publishing six inch to the mile maps. Meanwhile there remained an active market for civil engineers and surveyors and an instructive example is the life and work of William Bald (c1789-1857) recounted by Storrie (1969). Several engineers deserve notice for their contributions to the early study of coastal geomorphology: H.R.Palmer (1834), Sandford Fleming (1852), and the Stevenson firm based in Edinburgh and active throughout the nineteenth century.

Palmer's paper is accompanied by large scale plans of Dover and Folkestone Harbours, the former provided by the Admiralty, and these were supplemented by specially surveyed beach profiles. He concluded that benefits would accrue only if 'a system of management along the coast' were adopted, rather than 'particular devices adapted exclusively to each separate case.'

Sandford Fleming's independent coastal researches were became an entry to a competition promoted by the Commissioners of Toronto Harbour, who were worried by the progressive silting of the entrance. Fleming's entry, which won second prize, was prefaced by an essay on coastal geomorphology and accompanied by maps and surveys, including harbour soundings, made by Fleming himself at scales as large as twelve chains to the inch. His report is outstanding for its scientific content, but it was not alone. Two other civil engineers submitted reports, although the winner was Henry Y. Hind, Professor of Chemistry at Trinity College, Toronto.

The Stevenson firm in Edinburgh was noted especially for coastal, river and canal engineering. Robert Stevenson's 'River and Canal Engineering' is quoted by Lyell, and his father, Thomas, who specialized in coastal and lighthouse engineering, first discovered the subsurface wedge of salt water penetrating estuaries beneath the fresh river water, and described the enormous mechanical power of Atlantic storm waves in western Scotland.

Field Methods

It is appropriate at this juncture to consider the field methods used by the geologists of that period. Map creation is so slow that it is convenient, whenever possible, to plot field data onto the basis of existing maps. Where these are lacking, then, as the last section showed, precise surveys may be made for special purposes. The need to record field observations suggested itself to Woodward, though apparently no field notebooks survive. Wedgewood, in the letter quoted, provided a rough drawing of the strata encountered in the canal section, and Hutton appended geological sections of unconformities to his published volumes. Regrettably, for the progress of geology, his volume of drawings were lost, and have only recently been published (Hutton 1978).

A paper by Webb Seymour (1815), describing fieldwork undertaken in Glen Tilt by himself and John Playfair, provides information on field techniques. A footnote (p305) describes a clinometer 'in its early form, without the brass plate' used for taking dips and assumed to be accurate 'within a few degrees of the truth.' This was used with a compass since the bearing of the stretch (strike) is given. Points of interest were plotted on the best available map, Stobie's map of Perthshire, and an attempt was given 'to give a better representation of the mountains near Glen Tilt.' For fine details a map of the bed of the Tilt was constructed 'by taking bearings with the compass along the bank of the river; according to its changes of direction, and by pacing the distances. The length of a certain number of paces was determined by measurement.' Later it is noted that in some of these plans 'the rocks were merely sketched in by eye.' In this manner nearly two miles of the Tilt were mapped, and published at the scale of 1:7200, with detailed plans of four limited areas at one inch to fifty feet. Field measurement errors are mentioned for one detailed plan, but the point at issue is now obscure.

Absolute altitudes provided some problems. Barometrical methods were used by Playfair to refer the primary field site at Forest Lodge to the height of the lawn at Blair, fixed by General Roy at 425 feet above sea level. However, Webb Seymour notes that 'at the time of these observations, the barometer was falling, and the result is consequently too great.' When sites were inter-visible, Playfair preferred to rely on trigonometry and simple levelling:

> This is the highest point that is visible from the Glen below, and, by comparing its level with another on the same face, immediately above the Lodge, the height of which was afterwards determined by trigonometrical measurement, he computed it to be about eight hundred feet above the Tilt (Webb Seymour 1815).

Reconnaissance field methods have not changed a great deal in over one hundred and fifty years!

For a record of the landscape, and in the absence of cameras, a great deal of reliance was placed on the ability to make field sketches which were then engraved, usually professionally, and with various degrees of skill and success, to produce the finished plate. Hutton was fortunate to have a talented friend, John Clerk. Later, Lyell and Murchison in England, and James Hall, the New York State geologist, all possessed wives with that almost mandatory nineteenth-century skill - drawing. More accurate, if less artistic, was the camera lucida. Though not widely used, Captain Basil Hall (1788-1844, second son of Hutton's friend Sir James Hall) made a series of sketches of Niagara Falls (Hall 1829) that enabled Gilbert (1907) to detect an error in the first trigonometrical survey of the Falls (Hall 1843)! Photographs were not widely used in journals or texts until the last years of the nineteenth century and until then engravings, woodcuts, and lithographs held sway.

By 1850, field methods were routine and well established, and the British Admiralty issued 'A Manual of Scientific Enquiry: Prepared for the use of Her Majesty's Navy, and adapted to Travellers in general' (Herschel 1849). Charles Darwin wrote the section on geology. He described the use of the compass, clinometer and mountain barometer and went on to suggest that:

> a portable level would, in the case of raised beaches and terraces, be useful. Messrs. Adie and Son of Edinburgh, sell a land-level, a foot in length which is fitted with a little mirror on a hinge, so that the observer, whilst looking along the level, can see when the bubble of air is central, and thus instantly find his level in the surrounding district. Mr Chambers, moreover, and others have found, that an observer having previously ascertained the exact height of his eyes when standing upright, can measure the altitude of any point

with surprising accuracy. He has only to mark by the
level a recognizable stone or plant, and then to walk to
it, repeat the process, and keep an account of how many
times the levelling has been repeated in ascending to
the point, the height of which he wishes to ascertain.

Darwin makes particular mention of the use of the level for
determining the lateral inclination of raised beaches. The Mr
Chambers he mentions was a geologist who published the first
book devoted to raised beaches and terraces (Chambers 1848)
and which contains many surveyed measurements.

There is nothing surprising to modern eyes in the field
techniques used from 1800 onwards, but they developed in
response to the demands placed upon geologists by field
problems, and within the available technology. A recognition
of their intrinsic quality, even when their use is not
mentioned implicity, provides a measure of confidence when
reading the associated literature.

SCIENCE AND SOCIETY

Scientific societies and journals

The institutions and bureaucracy of science are so familiar
now that their very limited existence two centuries ago comes
as something of a shock. The Royal Society was founded in
1662, but there was no other national scientific society in
Britain until the Royal Society of Edinburgh received a Royal
Charter in 1783, although an embryonic philosophical society
had existed sporadically for several decades. The Linnean
Society was formed in 1788, with the active support of the
Royal Society, but the formation of the Geological Society in
1807 encountered initial hostility from the same body. The
Royal Society was originally set up to help popularize
science and to apply it to practical problems, but such
laudable aims were soon lost sight of. In 1799 the Royal
Institution was founded with similar aims and although it too
fell short of its intentions, it did (and still does) provide
lectures in science for the lay audience. Humphry Davy, the
chemist who invented the miner's safety lamp, gave a course
of geological lectures in 1805 and subsequent years, the
manuscripts of which have been preserved (Siegfried and Dott
1980). The most remarkable of the eighteenth-century
societies was the Lunar Society of Birmingham (Schofield
1963, Ritchie-Calder 1982), which although never formally
constituted, had an extra- ordinary influence on Midland
industry in the period 1760 to 1790, and numbered very many
Fellows of the Royal Society amongst its 'members.'

Most European countries also had their national
academies by 1800 (James Hutton belonged to the Royal Academy

of Agriculture in Paris), and they all functioned as clearing houses for developing scientific specialities. Through meetings, lectures and publications they provided a forum for discussion and were largely responsible for the structure of modern scientific society. The Geological Society of London was expressly founded 'for the purpose of making geologists acquainted with each other, of stimulating their zeal, of inducing them to adopt one nomenclature, of facilitating the communication of new facts.' Both Jameson (the Wernerian geologist at Edinburgh) and Playfair were elected honorary members, an honour only extended to non-residents of London. Although the financial demands of membership kept geologists such as William Smith and Robert Bakewell from being members, publication in the volumes of Transactions was not restricted to members. The Society emphasized the collection of facts and for a long time avoided formal theoretical discussions on the merits of the Wernerian or Huttonian schools of geology; a factor that may have inhibited, to a degree, the growth of geomorphological ideas.

Later in the nineteenth century, scientific societies, local and national, flourished and catered for all levels of competence. Geology was especially suited to local societies since every locale has different lessons to teach about earth history, and for a growing urban and industrial age, the pleasures of instruction in a rural setting made accessible by railways, were manifest.

The creation of scientific societies, as well as organizing scientists, helped formalize discussions about controversial topics and forced an explicit recognition of the methodology of science. All societies published volumes of papers read before their meetings, and, although often lengthy, they were a far cry from the voluminous books used to announce new theories, and typical of earlier centuries. Hutton's 'Theory of the Earth' was unusual: it was published both as a journal article (1788), and as a book (1795).

Theoretical views still found a way to surface in the addresses of society Presidents, men usually chosen in the first instance for their eminence in the subject. They rarely failed to take advantage of their opportunity to prosecute some favoured hypothesis. Lyell, reacting to Greenough's scepticism over the slow rise of Scandinavia out of the sea, 'found it particularly galling that such an incoherent series of theoretical criticisms should acquire great weight simply by their being uttered from the presidential chair of the Geological Society' (Wilson 1972). Lyell, in his turn, discussed the matter in his Presidential address of 1837, although he had already responded verbally.

One very important scientific forum of the nineteenth century was the British Association for the Advancement of Science, which was founded in 1830 and met annually thereafter. Organized into several sections, the Section

President was a leader in his field and if there was one date in the year that scientists reserved, it was that of the Annual Meeting. The Association was, and still is, famous for mediating between professional and popular audiences and provided scientists with a public platform largely lacking in their specialist societies.

Some journals led a life independent of scientific societies. A particularly good example is 'Silliman's Journal.' Silliman (1779-1864), an American lawyer, trained in chemistry and geology at Philadelphia and Edinburgh during the first decade of the nineteenth century. In Edinburgh he heard lectures by Playfair on the Huttonian Theory, and by Murray on the rival Wernerian Theory. Silliman became a professor at Yale College in 1802, although he did not lecture until 1806, after his return from Edinburgh. He founded his journal in 1818 as an outlet for arts and sciences in America. It still lives on as the 'American Journal of Science,' although its content is now primarily geochemical with occasional papers in geomorphology.

Another example is the 'New Edinburgh Philosophical Journal,' founded in 1826, and which was edited by Jameson who had trained with Werner in Freiburg, and whose lectures on geology at Edinburgh, in 1826, appalled Charles Darwin. Despite his theoretical views, Jameson allowed the journal to act as an open forum for all geologists, and his reprinting of an early notice of glaciation in Scandinavia (Esmark 1827) is a tribute to his tolerance. A decade later he reprinted important papers on glaciation by Agassiz and Charpentier.

Textbooks

A number of factors combined to ensure that the nineteenth century was one of self-improvement through education. The rising urban and industrial population created a new market for books of all kinds. Radical changes in printing technology; the application of steam to the printing press, the development of wood pulp paper, the use of lithography for cheap illustrations and the adoption of cloth book bindings instead of leather all permitted the production of much cheaper books after 1820 (Steinberg 1966).

The senior scientific societies formed the spearhead, but below them lay a market of thousands thirsting for instruction in all branches of science and natural history. Lyell initially intended to write a 'Conversations on Geology' for popular consumption before settling on the 'Principles of Geology' (1830-3) as a better means of expressing his philosophy. Nevertheless, in 1828 he invited his friend Gideon Mantell, the Sussex paleontologist, surgeon and popular geologist, to fulfill this need, which he eventually did with his 'Thoughts of a Pebble' (1836) and 'Wonders of Geology' (1838); both ran to many editions.

One of the earliest formal textbooks in geology was Bakewell's 'Introduction to Geology' (1813) which went through five English editions (5th ed. 1838), a German translation, and three American editions, by courtesy of the efforts of Benjamin Silliman, who adopted it as his course textbook. It was, perhaps, the most moderate and reliable of the early geology texts, and was 'intended to convey a practical knowledge of the science.' It is credited with introducing Charles Lyell to geology through a copy of the second edition in his father's library. It steered a middle road between the extreme camps on almost all topics, and is full of sound practical geology, ranging from the provision of what he claimed was the first published map of the geology of England and Wales to the observation that limestone fissures are partly due to solutional processes.

The main geological textbook of the nineteenth century was Lyell's 'Principles of Geology,' which eventually went through twelve editions. A reduced version, essentially concerned with the stratigraphy and regional geology of Britain, was produced from 1838 onwards entitled 'The Elements of Geology.' Later in the century even more elementary abstractions such as 'The Student's Lyell,' served the lowest level, and innumerable students got their first taste of geology from one of Lyell's texts, on both sides of the Atlantic. In North America, Dana's 'Manual of Geology' (1st edition 1863) dominated the market after its issue, but in Britain there was no serious competitor until Geikie's 'Textbook of Geology' appeared in 1882, following Lyell's death in 1875.

A concern with surface features occupied progressively less space in geology texts as the century went by. However, a lay interest in scenery persisted and the response was a series of books explaining regional geology and surface features. Typical and influential examples were Ramsay's 'Physical Geology and Geography of Great Britain' (1863), which grew out of a popular lecture series, Geikie's 'Scenery of Scotland' (1865), and Hull's 'Physical Geology and Geography of Ireland' (1878).

Scientific travellers

In an age of television, documentaries and space travel, it is difficult now to imagine the dependence of geological theorists around 1800 on reports of foreign travels. Playfair recorded Hutton's avid reading of voyages, travels and accounts of natural history, and Playfair's own library contained a full selection of such volumes.

Three significant eighteenth-century figures were Simon Pallas (1741-1811), Horace B. de Saussure (1740-1799), and Jean Andre de Luc (1727-1817). Pallas was a German, attached to the Russian Court in St Petersburg, who described travels made eastwards into Siberia during the years 1768 to 1774. He is famous for his records of the frozen fauna; elephant, rhinoceros and mammoth finds that caused much puzzlement in Europe and confirmed in Sir James Hall his belief that huge floods had swept north from the tropics.

Saussure was a Swiss geologist who published four volumes of travels recording explorations in the Alps, including an ascent of Mt Blanc. He wrote that 'it is the study of mountains which above all can quicken the progress of the theory of the earth or geology.' The first two volumes made a profound impresssion on Hutton, who quarried them for material in support of his 'Theory,' and his debt to Saussure was incalculable. Saussure's clear descriptions of glaciation led both Hutton and Playfair to extend the glaciers outwards sixty miles from the Mt Blanc massif to explain the Jura erratics. As Geikie (1905) remarked, 'it was under the guidance of the great Swiss observer that the Scottish philosopher stood in imagination on the summit of the Alps.'

Hutton also relied on the descriptions of de Luc: 'science is indebted to this author for giving us so clear a picture of natural appearances, and of his own reasoning upon those facts, in forming his opinions; he thus leads astray no person of sound judgement, although he may be in error' (I-379). Now wonder de Luc responded so vigorously to Hutton's interpretations of sites he had seen only through the eyes of de Luc and Saussure!

Two great German travellers provided a similar service to the first half of the nineteenth century: Leopold von Buch (1774-1853) was primarily a geologist who confined himself to Europe, whereas Baron von Humboldt travelled the world as a natural scientist. It would be difficult to cite a serious nineteenth-century text that did not make reference to one of these great figures. Their theoretical views were often askance; von Buch, for example, believed that volcanic mountains were elevated from beneath, and both objected strongly to the glacial theory. Their great personal authority imposed diluvial conservatism on German geology. However, failing a personal visit to a field site, a description by Humboldt or von Buch was the next best thing.

Another type of scientific traveller was the natural scientist attached to an official expedition, usually financed by the Admiralty. The most famous example is Charles Darwin, on the Beagle with Captain Fitzroy. Darwin produced much valuable geology in addition to his botany and zoology, and two items of particular importance were his study of coral reefs, and his reports on coastal elevation and submergence along the South American coast, and their relation to earthquakes. Both these topics made excellent material for Lyell, and they figured prominently in later editions of the 'Principles.' His reports of boulders lodged on icebergs in the South Atlantic also impressed Lyell, who remained long in the grip of an hypothesis of glacial submergence, with glacial drift delivered by icebergs.

Other voyager-geologists of the nineteenth century were James Dana of the United States, who made studies of coral reefs and volcanos (1849), and J.B Jukes who gave the first scientific description of the Great Barrier Reef, in addition to other western Pacific corals (1847). Arctic exploration yielded valuable material in the hands of meticulous naval captains (Scoresby, Bayfield, Harcourt), and the travelling amateur geologist, for example Lyell's friend Basil Hall, remained a trusted source well into the nineteenth century.

A similar opportunity to the Darwin voyage which might have had an important impact on geology in the 1780's, was lost when Joseph Banks, later President of the Royal Society for forty two years, left his notebooks unpublished after his voyages as a naturalist with Captain Cook in the Pacific. Late in the century the CHALLENGER expedition (1873-6) added greatly to geologists' knowledge of the ocean floor.

IN CONCLUSION

The public face of science underwent such a radical change between 1790 and 1830 that it is easy to forget the continuity of the underlying private practice. Geology, perhaps more than most subjects, was transformed, and with the appearance of Lyell's 'Principles of Geology' (1830-3) the subject was placed on thoroughly modern foundations. But, with the establishment of the discipline, the clear statement of its component parts, and the impetus of industrial expansion in the nineteenth century, geology found that its principal interest lay beneath the ground, with a temporal perspective going back in time to the origins of rocks and the beginning of time. Geomorphology, in contrast, had served its purpose to geology by scratching the surface to show what lay below, and could not help but look forward in time, from a shadowy birth as an unshaped mass to a present splendid sculpture. It was to be left behind, like a hapless rural milkmaid at the pit head, as the miner climbed below.

Chapter Five

THE PROGRESS OF UNIFORMITARIANISM

Chaos and confusion are not to be introduced into the order of nature, because certain things appear to our partial views as being in some disorder.

(James Hutton, 1795, II-547)

The period from 1820 to about 1880 is a very confused one in the history of geomorphology. The subject was slowly disentangling itself from the main concerns of geology and would not establish its identity until the end of the century. The next three chapters are concerned with three main threads that help render this period understandable: the progress of uniformitarianism, especially fluvialism, the metamorphosis of catastrophism, particularly its interest in floods, and the idea of glaciation which would eventually resolve crucial problems for both schools of thought. In the following chapters, it is impossible, and in any case undesirable, to maintain a total purity of theme; inevitably there will be an overspill, especially as many of the actors appear on all three stages.

The pace of field investigation during the nineteenth century was such that a brief account can only mention a few names in passing. To provide some coherence I have tried to illustrate the main ideas with particular case studies. Davies (1969) notes that for Victorians, resolving the geological and Mosaic chronologies was a hobby, but the voluminous writings of the 'scriptural geologists' had no serious impacts on the development of the subject.

In order to understand some of the conceptual problems faced by the early nineteenth-century geologist, it is essential for the modern reader to discard, in his imagination, the Quaternary Era. He must then try to visualize the development of the geological column, and the landscape developed on its rocks, smoothly up to the present time, i.e. precisely the type of mental gymnastics that martian and venusian planetary scientists of the future may have to undergo. Of course, those who clung to Mosaic chronologies had, perforce, to introduce the Biblical Flood, at about four thousand years ago, into the scheme of things. However, strict scripturalists were soon a thing of the past, and for serious geologists the logical position, within the general theory of geology adopted loosely from Playfair's account of the Huttonian Theory, called for the continuous operation of processes, at least since the last continental upheaval, right down to the present.

We now know that the recent geological past has been characterized by radical changes in the behaviour of almost all the surface processes of which we have knowledge. Succinctly, Pleistocene glaciation had systemic effects on the whole globe. The field evidence for this is well known, and extremely diverse in character. At the very least it provides even the casual observer with evidence that contradicts any notion that the present surface has developed by essentially present, and local, processes since the Tertiary. It was the interpretation of this field evidence that caused the ideological differences between the rival schools of uniformitarians and catastrophists up until the 1870's. There was much less dispute about the field evidence itself; the facts were largely admitted.

CHARLES LYELL AND 'THE PRINCIPLES OF GEOLOGY' (1830-1833)

'The uniformity of present laws'

Charles Lyell (1797-1875) dominated nineteenth-century geology as no man has dominated geology since. He trained, and practised briefly, as a lawyer, but a natural inclination to geology and problems with eyestrain persuaded him to try to earn a living from his writings and lecturing as a geologist. By living modestly he was able to travel widely at home and abroad, and by the time he published the first volume of the 'Principles of Geology' in 1830 he was already one of the best known British geologists. He had corresponded with, or visited, almost all continental geologists of note, and the footnotes of the 'Principles,' in all its various editions, provide a fascinating record of his sphere of influence. More than one might expect, the 'Principles' is a record of European, not merely British, geological thinking. In quoting from Lyell's 'Principles' I shall use the convention: date, edition, volume, page(s), thus: 1830-1-i-144.

The book is subtitled 'An attempt to explain the former changes of the earth's surface by reference to causes now in operation.' The first volume carries a quotation from Playfair expressing the fundamental notion of what Lyell termed the 'uniformity of physical laws.' This implied 'that all former changes of the organic and inorganic creation are referrible to one uninterrupted succession of physical events, governed by the laws now in operation' (1830-1-i-144). The organic changes are of no immediate concern here, although, as the years went by, concerted opposition to Lyell's principle of uniformity was orchestrated by those who sought a progression in the development of the organic world. This problem was thrown into sharp relief when Charles Darwin, perhaps Lyell's closest scientific friend, published 'On the Origin of Species' (1859), and it caused Lyell much soul-searching to accommodate his scheme to Darwinism.

Lyell was wise enough to be very careful about how the principle of uniformity might have to be interpreted, at least with respect to inorganic events:

> But should we ever establish by unequivocal proofs, that certain agents have, at particular periods of past time, been more potent instruments of changes over the entire surface of the earth than they are now, it will be more consistent with philosophical caution to presume, that after an interval of quiescence they will recover their pristine vigour, than to regard them as worn out (1830-1-i-165).

The careful distinctions that Lyell was trying to draw were easily lost, even on sympathetic critics. This is highlighted by a remark in the third volume of the 'Principles':

> We regret, however, to find that the bearing of our arguments in the first volume has been misunderstood.. for we have been charged with endeavouring to establish the proposition, that 'the existing causes of change have operated with absolute uniformity from all eternity' (1833-1-iii-383).

The criticism appeared in a review article, specially commissioned for the 'Quarterly Review' and written by Scrope (1830), 'a friendly critic,' known for his work on the volcanos of central France. Lyell, however, had merely wished to establish, like Hutton and Playfair before him, and the Ancients before them, that:

> - the value of all geological evidence..must depend entirely on the degree of confidence which we feel in regard to the permanency of the laws of nature. Their immutable constancy alone enables us to reason from analogy, by the strictest rules of induction, respecting the events of former ages..(1830-1-i-165).

Misunderstanding was still possible, as Adam Sedgwick, the Woodwardian Professor of Geology at Cambridge, demonstrated in his Presidential address to the Geological Society in 1831 when he wrote that 'the earth's surface ought to present an indefinite succession of similar phenomena' if Lyell's doctrines were true. Rather more acute was the criticism of William Whewell. He reviewed the now abundant field evidence and concluded that 'they spoke of a break in the continuity of nature's operations; of the present state of things as permanent and tranquil, the past having been progressive and violent' (Whewell 1831). Clearly, the problem lay, as noted above, with interpreting the field evidence.

The potential American reaction is revealed when Lyell wrote in a letter (1831) to his friend Mantell: 'Yet Featherstonehaugh tells Murchison in a letter, that in the United States he should hardly dare in a review to approve of my doctrines, such a storm would the orthodox raise against him.' On the continent the position was little better. In Germany, the powerful influence of von Buch and Humboldt, both catastrophists, was little modified by the academic work of von Hoff (1824-1834) who tended toward Lyellian views. The catastrophic bias in Germany lasted until the work of Rutimeyer (1869) on valley and lake formation in the Alps.

In France by 1830, the position had become even more conservative and marked a radical shift from the fluvial and uniformitarian school of the previous century. Cuvier and Elie de Beaumont, paleontologist and structural geologist respectively, held extreme catastrophic views on the origin of fossils and mountain building. Cuvier long supported the idea of the Mosaic Deluge, and Beaumont thought that whole mountain chains were formed in catastrophic upheavals. A book by Lamarck on 'Hydrogeologie' (1802, privately published in a small edition, Eng. trs. 1960), with gradualist principles was the last remnant of the fluvial school and it contained a curiously 'reversed' erosion cycle. Even so this, the translation of Playfair's 'Illustrations' (Basset 1815), and the pioneering work of the continental hydrodynamicists of the eighteenth century, culminating with Du Buat (1779), were of little avail in promoting uniformitarian principles and Lyell frequently complained that French geologists were too pre-occupied with politics.

The sources of Lyell's uniformitarianism are not entirely clear. His first acquaintance with geology was Bakewell's 'Introduction to Geology' (1815 2nd ed.), which, although fair to Hutton and Playfair, maintained diluvial attitudes in all editions. At Oxford (1816-21) he attended the lectures of William Buckland who remained a staunch diluvialist until his conversion to the glacial theory in 1840. In 1823 he went to France to improve his French and 'geologized' with Constant Prevost whose approach, in mapping the Paris basin, was philosophically uniformitarian. Prevost was a friend of Dr Fitton, who was later to accuse Lyell of understating his debt to Hutton (Fitton 1839). While in France he probably encountered the work of the fluvial school focussed on the Auvergne region, and he certainly met the aging Montlosier. These connections and the reprinting of Playfair's 'Illustrations' (Playfair 1822), probably go some way toward explaining Lyell's adoption of the notion of uniformity, but this should not be confused with the more restricted and specialized idea of fluvialism.

The initial debate about uniformity

The appearance of the 'Principles' polarized philosophical opinions about geology, and the ideological positions adopted were not to be readily abandonned. Of equal importance, the 'Principles' provided a storehouse of anecdotal information on erosional phenomena, and on the manifestations of renovating forces such as volcanos, earthquakes and imperceptible uplift, or sinking, of the land surface. Part of Lyell's aim was to provide a collection of observations which could serve to define 'the succession of physical events governed by laws now in operation' that he regarded as fundamental to his system. He went to considerable lengths to show that even at present, which he took as the historical past of about 3000 years, the magnitude of events is not distributed evenly in space or time. An exhaustive inventory of known earthquakes established both their ubiquity and their usual results: vertical motion of very substantial sections of the earth's crust. In general, he was able to show that the net movement was upwards, and to reduce the magnitudes to an imaginable scale, he used a unit of measure equal to the Great Pyramid! He also compared the work done by an earthquake in a few hours to that done by a river like the Mississippi, or the Ganges, in a year. The Ganges, he computed, removes 60 pyramids a year, and would require over seventeen centuries to remove the mass uplifted in a single Chilean earthquake. The average amount by which an earthquake uplifts land is shown to be the order of just a few feet, but he did note the example of the Lisbon earthquake in which the sea bed close the quay was reputedly depressed by six hundred feet. With examples of this sort Lyell could encompass within the uniformitarian system quite substantial 'catastrophes.'

In fact the debate about uniformity, in the decade after 1830, centred rather more on the issue of valleys, and whether they were excavated by rivers, than on the behaviour of volcanos and earthquakes, and about which there could be little debate about the meaning of the field evidence. Because Lyell is universally associated with nineteenth-century uniformitarianism, and because in large part this position agrees with the views of Hutton and Playfair, he has often been criticized for failing to support the fluvial school to the full (Chorley et al. 1964, Cunningham 1977a,b). However, nowhere in the 'Principles,' or anywhere else, does Lyell state that he is adopting and defending Huttonian Theory in its entirety, and it would appear that the criticism is, to some extent, misplaced. The origin of valleys was still a topic of heated discussion in 1830 because Lyell and Murchison had, the year before, read a paper to the Geological Society based on field work in

Auvergne, central France. In this region, as Scrope had already shown in 1827, like Guettard and Desmarest in the century before, volcanism repeatedly filled up river valleys with basalt lava flows, just as often to be re-excavated by river action. Unmistakable river gravels are preserved below the valley-filling lavas and the efficacy of fluvial action was indisputable. However, evidence can always be interpreted in at least two ways! The catastrophists, bolstered by a paper from Coneybeare on the formation of the Thames valley and usually favouring a single Mosaic flood, were driven to a series of floods, one for each preserved river valley, or in the Thames valley, river terrace, and the recognition that no flood had occurred since to remove the fragile cinder cones in central France, or on Mt Etna in Sicily. They did not regard the evidence as positively in favour of efficient fluvial action. Lyell, who was a fluvialist up to a point, saw the evidence for what it was, and he wrote to Mantell that 'Coneybeare..admits 3 deluges before the Noachian! & Buckland adds God knows how many catastrophes besides so we have driven them out of the Mosaic record fairly' (Wilson 1972). However, Murchison, as he recorded later in his 'Silurian System' (1839), merely saw the case as:

> among the first which fairly brought the diluvial question to issue in this country, and gave rise to those discussions which led to the refutation of the belief in a general terrestial deluge, which had affected simultaneously all the surface of the earth.

Murchison never subsequently shifted from a multiple deluge position, and never ascribed significant erosion in hard rocks to modern rivers. Coneybeare (1831), for his part, put the diluvial position very plainly:

> Do I then deny that fluvial erosion has ever produced a single valley? and if so, how do I dispose of the evidence..in favour of this view? I will avow the tendency of my arguments openly and frankly. I deny that..any..valleys of excavation..have been so produced, except under extraordinary circumstances. And to the evidence I reply that it relates to districts in which these extraordinary circumstances undoubtedly exist, -volcanic districts for instance, such as Auvergne and Etna. Now I cannot admit the action of torrents occasioned by, and cooperating with, volcanic convulsions, as an example of the ordinary action of common streams; -but that under these extraordinary circumstances, and even under such more common but still comparatively rare incidents as the late floods in Scotland, fluvial action may occasionally produce considerable effects, I do not deny.

No wonder that Lyell thought that there was more than 'an imaginary line of separation between' him and the catastrophists, and that the matter was more than 'a dispute about degree, a plus or minus affair' as Scrope had written to him in private (1832). Nevertheless, as time went by, the need for multiple floods in the catastrophist school, and the great variety and magnitude of physical events described and envisioned by the uniformitarians, did force the schools closer together. After 1840, the appearance of the glacial theory provided yet another complicating factor, some aspects of which were adopted by both schools of thought.

Coneybeare, writing in 1832, thought that since 1821 'geology has received scarcely any valuable addition, and not a single fundamental one' (Wilson 1972). The opinion under-valued, if it did not ignore, the principle of uniformity, and it failed to appreciate, not surprisingly, the conceptual significance of a shift away from a single flood. Never-theless, as a result of the discussions on the 'Principles' and the Auvergne valleys, Sedgwick, De la Beche, Greenough, and even Buckland, all, in time, shifted away from a single, to a multiple flood, position. The convergence of the two lines of thought, mentioned by Scrope, becomes more evident when the more extreme aspects of Lyell's 'uniformity' are examined in the next section.

'The succession of physical events'

Lyell, as the proponent of uniformity, is frequently down-played in his advocacy of relatively catastrophic events. To avoid misunderstanding I shall state that all Lyell's extreme events are carefully documented, or follow from field evidence by the application of logical geological principles. In the latter class are his speculations on the destruction which would accompany the final retreat of Niagara Falls into Lake Erie and the flood that this would precipitate onto the country below. In addition to the Lisbon earthquake, mentioned above, he documents the one that occurred in the Mississippi valley at New Madrid in 1812, and still the largest historical earthquake on record in the United States, with the consequent creation of islands in the river, and lakes on the flood plain.

Deluges that occur when lake barriers break are treated at some length in the 'Principles.' A consideration of such floods leads Lyell to regard such events as commonplace in the development of river valleys:

> The power which running water may exert, in the lapse of ages, in widening and deepening a valley, does not so much depend on the volume and velocity of the stream usually flowing in it, as on the number and magnitude of the obstructions which have, at different periods,

opposed its free passage (1830-1-i-192).

When this idea is expressed more generally it sounds like an approach to the magnitude/frequency concept developed in the 1960s (p188):

> It is evident, therefore, that when we are speculating on the excavating force which running water may have exerted in a particular valley, the most important question is not the volume of the existing stream, nor the present levels of the river-channel, nor the size of the gravel, but the probability of a succession of floods, at some period since the time when some of the land in question may have been first elevated above the bottom of the sea (1830-1-i-196).

Lyell is impressed with such happenings, and sees them less as part of the ordinary fluvial behaviour, than as mild catastrophes imposed by tectonic forces, with their side effects, landslips creating dammed lakes. From this perspective it is clear that the quotation is not even hinting at the magnitude/frequency concept, although it may presage an ultra-modern neocatastrophist view! He specifically links hydrographic changes in large valleys, like the Mississippi, with the convulsions of earthquakes, and related volcanic eruptions. He suggests that 'some theorists .. might no longer feel themselves under the necessity of resorting to catastrophes out of the ordinary course of nature to explain the alluvial phenomena of that district.' In this quotation he was using the New Madrid earthquake as exemplar for what might have happened in Auvergne, or the Eifel, a similar volcanic district.

Lyell never altered his stance on the principle of uniformity, but this should not be confused with alterations of his views on certain categories of phenomena. The slow uplift of Scandinavia, mentioned by Playfair (1802 #391), is one example of this. In the first edition, Lyell remarked that 'no countries have been more free from earthquakes since the time of authentic history than Norway, Sweden and Denmark' (1830-1-i-231). He therefore doubted the 'extraordinary notion' of 'Von Buch, who imagines that the whole of the land along the northern and western shores of the Baltic is slowly and insensibly rising!' The matter was so important, and the reports so persistent that Lyell made a special trip in 1834 to investigate, and by the fifth edition it was incorporated into his thinking:

> The fact also of a very gradual and insensible elevation of the land may explain many geological monuments of elevation, on a grand scale....How easily may oceanic currents..sweep away this thin layer of matter thus

brought up annually within the sphere of aqueous
denudation (1837-5-ii-305).

Thus, to the inventory of local catastrophes is added the
perfect complement: uplift of insensible continuity.
Likewise, Darwin's (1838) theory of the formation of coral
reefs suggested a slow subsidence of the ocean bottom in the
Pacific: precisely the opposite effect. In the light of these
facts it is little wonder that Lyell subsequently put
considerable emphasis on the efficiency of the sea as an
agent of erosion, particularly as a landmass rose gradually
out of the ocean. The Scandinavian example was heaven-sent to
a uniformitarian surrounded by catastrophists and
diluvialists. He applied the idea in particular to the
landforms of the Weald, whose inward facing escarpments he
attributed to marine erosion as the Wealden dome emerged
slowly from the sea. The structure of the Weald was a good
example of a phenomenon that nineteenth-century geologists
termed a 'denudation' i.e. the word was used as a noun.

Denudations

The amount of erosion which had to be accomplished to remove
all the volume of rock indicated by an uninterrupted
projection of existing strata dawned slowly on geologists. To
a catastrophic theory, the amount removed by a diluvial wave
was relatively immaterial (and the term 'denudation' carried
catastrophic overtones), but in a uniformitarian perspective
it implied enormous amounts of elapsed time. Hutton had
realized this (Chapter Three, p51) but the reassertion of
uniformity, and a shift to multiple deluges, placed a new
significance on the volume of rock lost. In Chapter Four,
Bakewell was revealed chastizing himself for not emphasizing
the existence of huge denudations in coal-field regions; an
echo of Hutton's revelation of a 'slip, or hitch..of 70
fathoms..the surface on each side of this line is perfectly
equal' (1795, II-289). In the case of the Weald, Coneybeare
thought 'it would be highly rash to assume that the chalk at
any period actually covered the whole space,' but Lyell found
himself compelled to accept that it had. The issue,
therefore, was on the mechanism of excavation. Lyell thought
that this was the sea, working on longitudinal and cross
fissures, which therefore produced the transverse valleys of
the Weald. In support of this conjecture he quoted and
illustrated a valley (the Coomb, near Lewes in Sussex) which
is associated with a fault (1833-1-iii-301).

Most nineteenth-century discussion of denudations took
place with reference to structural units such as the Wealden
anticlinorium or the Lake District. This arose from adherence
to Huttonian theory: rocks form below the sea, and as they
are uplifted above it the structural forms develop, and

fissures and fractures then dictate the way in which either the sea, or subaerial erosion, develops the river pattern. In this way both diastrophism and erosion were intimately linked and, to a considerable extent, could develop independently in different parts of the country: 'We are aware that we cannot generalize these views and apply them to the valleys of all parts of the earth' (1833-1-iii-319). Clearly, uniformity applied only to the way in which events could be expected to operate; it did not select or suggest in a particular circumstance which series of events had actually taken place.

The role of weathering

The slow reduction of solid rock to a particulate state by physical and chemical weathering before its removal by erosional agents is so important that it might be expected to be a cornerstone of the uniformitarian edifice. However, this was not so. Weathering was assumed as an act of faith, but the details of the processes were not studied. For the catastrophists weathering was irrelevant to the creation of topography since it was a contemporary process, and after 1830 its existence was not denied. To the uniformitarians the fundamental issues were much larger; which were the more important processes responsible for shaping the surface: fluvial or marine, and how did these operate in space and time to remove loose material? The environmental bias of high latitudes had a part to play here since, by and large, Holocene weathering and soil formation has not been dramatic, and the troublesome problem of superficial drift loomed much larger in the explanation of surface topography.

Consequently, the processes of weathering were not seen as crucial by either of the contending schools of geologists, and this is reflected in their writings. This is not say that the main processes were not appreciated. Hutton described freeze-thaw, the solution of limestone (both following Saussure 1779) and the decomposition of the feldspar in granite to kaolin. Playfair (#341) described a frost-shattered mountain summit and posited a steady-state theory of soil formation such that 'the supply and waste of soil are exactly equal to one another' (#103). In contrast, Lyell completely neglects weathering processes and referred readers of his first edition to Bakewell on the subject of soil formation. This neglect persisted in all subsequent editions of the 'Principles' and thus the first systematic accounts of weathering are to be found in the second generation of textbooks such as Dana (1863) and Geikie (1882) under the heading of 'dynamical geology.'

Individual papers on weathering are relatively rare although notable early ones are Yates (1830-1) who noted the importance of oxidation, Phillips (1831) who drew attention to cycles of moisture and heat in causing exfoliation of a rock surface independent of its internal structure, and Bartlett (1832) and Adie (1835), both of whom experimented on the thermal expansion of common building stones. Late nineteenth-century accounts naturally revealed the maturation expected from many decades of physics and chemistry stimulated by the Industrial Revolution, and comprehensive descriptions were given by Ansted (1871), Powell (1875) and Gilbert (1877). Gilbert remarked on the role of water as an almost universal solvent, on the propensity of sedimentary rocks to weather by the loosening of their cements, and on the chemical and physical action of plant life in rock decay. In the second half of the century explorers' reports made available a surprising wealth of anecdotal material on the importance of weathering in tropical regions but in the absence of prolonged field study, and a proper physiographic framework, no systematic progress was possible. Perhaps pride of place should go to Stanley's observation on the importance of thunderstorms to the sudden cooling of sun-heated rock surfaces (Stanley 1876).

Although weathering was never completely neglected it was not, nor was it seen to be crucial to a resolution of the dominating controversies of the century.

The principle of uniformity, 1835-1875

Lyell's 'Principles,' and their several derivatives, remained pre-eminent in Britain until Lyell's death in 1875. If the distinction between the abstract principle, its illustration by 'modern' examples, pure Huttonian theory, and pure fluvialism are all kept in mind, then it may be said the principle served its purpose, just as the 'Principles' served to illustrate it, by providing a powerful counterpoise to cataclysmic theorizing. After the initial debate in the 1830's, the situation resolved into debates on specific topics as each side sought field evidence, or developed theoretical analysis, to support its views. It did, as Scrope had sagaciously remarked, become a matter of degree. Paleontology, by mid-century, was established as the central concern of geology; Mantell noted with disgust in his diary (18th Oct. 1849) that a paper on the viscous theory of ice, by J.D.Forbes (a glacialist) was being considered for a Royal Medal of the Geological Society. As young blood decided for itself how it would reason, and upon what evidence, the categorization of individuals into particular camps becomes a merely semantic exercise, of little practical value.

The 'Principles' appeared in translation in German, French and Spanish, and in American editions. Lyell propagated his views in North America with lengthy trips in 1841-2 and 1846-7, on both of which he lectured to large audiences. If Featherstonehaugh was nervous, though accepting, of uniformity in 1830, James Hall, a young State geologist in

New York, had no such qualms in 1843, when he published the results of his work in the Fourth District of New York. Although the structure of his report was closely modelled on the 'Silurian System' of Murchison, its geomorphology was Lyellian but also fluvial: Hall had escorted Lyell in the field in 1841. A decade later, Sandford Fleming (1854) was writing a classic of coastal geomorphology from a perfectly modern standpoint (Gilbert 1890), and in 1863 James Dana published his 'Manual of Geology' which became the established American textbook for the remainder of the century.

By the latter half of the century, on both sides of the Atlantic, Hutton and Playfair were mere memories, and their publications were scarce. Geology as a whole had become so complex that 'uniformity' as such was not a debatable issue. Lip service was paid, but discussion concerned the evidence revealed in field sites and sections. Darwinism and paleontology diverted attention within geology; evolution became the catchword although uniformity had prepared the path. In the field of landforms, glacial theory was the diverting agent. The sheer mass of field evidence in both fields began to outweigh the usefulness of a simple philosophical principle: more comprehensive organizing frameworks were required.

The rejuvenation of the fluvial idea under the umbrella of uniformitarianism is of such abiding interest to geomorphology that the next section will look at this particular topic in more depth.

THE VICISSITUDES OF FLUVIALISM

In the debate about fluvialism, one subset of the uniformity debate, the valleys of central France formed one widely studied case example, and Coneybeare introduced others: the Vale of Kingsclere, the Thames Valley, and Niagara Falls, in his efforts to establish the non-fluvial origin of valleys. Niagara Falls might have acted as a banner for the fluvialist cause, but instead it acted as barometer, responding to the ever-changing pressures of current opinion.

It had the advantages of not calling upon the very special agent of volcanism, and of displaying in a thundering spectacle, the supposed agent of its creation. It had the disadvantage of lying far from the centres of advocacy in a dispute it might have resolved, and in addition it displayed particular features of its own that called for special pleading. Before I turn to a longitudinal study of thought on Niagara Falls I will try to illustrate, using the example of Lyell's 'Principles,' the general framework of geological thought on the activity of rivers during the first third of the nineteenth century, and the sources of hydrodynamical research from which it was drawn.

The facts of fluvialism

The hydrodynamical behaviour of water in flowing channels was comprehensively studied in the eighteenth century by Italian, French and Swiss civil engineers. Their results, couched in equations that still appear in textbooks, were available to any interested reader, and Playfair, for example, had a comprehensive collection in his library. Du Buat (1779, 1786, 1816), a French engineer, likened the stages of a river along its profile to 'the different ages of man' and depicted its beginning, infancy, youth, middle course, and old age. The long profile was described in these terms:

> The bed of rivers does not form a uniformly sloping plane but is the combination of several continuous inclined planes the gradients of which constantly decrease towards the sea (in Chorley et al. 1964, p89).

Du Buat discussed velocity as resulting from an equilibrium between the impelling force of slope and the resistance of the channel. By comparisons he then showed that stream bends introduced additional friction. These theoretical concepts were far in advance of any use to which geology could put them. A more practical use might have been made of the table showing the velocities which would entrain various types of sediment from clays to coarse gravels; this at a time when the transportation of sediment in streams was vigorously denied by De Luc and Williams, amongst others. In a similar vein Lamarck (1802) tried to establish the efficacy of rivers but the book made no mark, and scientific circles preferred his views on variation in species.

Lyell (1830) makes no mention of the continental civil engineers in his discussion of rivers but his presentation is sound and logical. He notes the vertical and horizontal velocity profiles in rivers, repeats Playfair's point that sediment immersed in water loses between a third and half of its specific gravity, and gives entrainment velocities for fine clay, fine sand, fine gravel, and egg-sized gravel. His information is derived from an article on 'Rivers'in the 'Encyclopedia Britannica' which was authored by John Robison, a colleague of Playfair's, sometime in the period 1779 to 1797. Robison drew heavily on the work of the continental engineers, and especially Du Buat.

Lyell then asks why, if streams can entrain all this sediment in their upper courses, their lower courses do not become clogged as the the slope lessens? His answer, again taken from the continental civil engineers via Robison's article, and neglected by later geomorphologists until rediscovered in the 1950's, was that:

> this evil is prevented by a general law regulating the

conduct of running water, that two equal streams do not occupy a bed of double surface. In proportion, therefore, as the whole fluid mass increases, the space which it occupies decreases relatively to the volume of the water; and hence there is a smaller proportion of the whole retarded by friction against the bottom and sides of the channel. The portion thus unimpeded moves with greater velocity, so that the main current is often accelerated in the lower country, notwithstanding that the slope of the channel is lessened. It not infrequently happens..that two large rivers, after their junction, have only the surface which one of them had previously.. By this beautiful adjustment, the water which drains the interior country is made to continually occupy less room as it approaches the sea; (1830-1-i-173).

Noteworthy in this extract is the statement of a general law; that the velocity increases downstream, that channel shape adjusts to increasing discharge, and that this is due to a relative reduction in friction. The whole constitutes a 'beautiful adjustment' for the system as a whole. However, as with the work of Du Buat, the work of the engineers fell on deaf ears, and although the statement remained in all editions of the 'Principles' most of its readers preferred anecdotal catastrophism. Lyell continued by illustrating the power of running water with reference to recent floods in Scotland, and then broadened the discussion to take in gullies eroded in volcanic materials on Vesuvius, and in Mexico, and the valleys gradually excavated out of resistant lavas in central France, on Etna in similar fashion, and finally the gorge of Niagara.

His next chapter has been noted above, and treats of mild catastrophes of various types that can affect river valleys; dam bursts and earthquakes, but which are not part of what we would regard as the normal hydrological regime.

From this review it is clear that a fairly broad and sound basis for understanding the behaviour of rivers was readily available by 1830, most of it originating from the work of civil engineers. The problem was, then as now, how does this knowledge bear on the issue of valley formation?

Objections to a fluvial origin for valleys

Objections to the idea that valleys are eroded entirely by the rivers that occupy them appeared as soon as the idea itself took serious hold. Indeed the objections serve to show that the issue was a live one, and taken seriously. Some of this reaction has been documented in Chapter Three. The objections raised were in general perfectly valid. Davies (1969) documents at least eleven, but six will suffice to cover the main issues. The points mainly arise, as we now recognize, from what can be termed the 'climatic accident' of Quaternary glaciation.

Perhaps the primary objection was the limnological problem. Fluvial theory maintained that the landscape eroded by slow downwearing. Lakes, if they ever existed, only could have done so early in the history of the landscape, and should have been rapidly filled in. If this was so, how was it that in deeply eroded landscapes, and in the course of very large valleys, very deep lakes could be found? Lake Geneva has already been discussed in this context, and Playfair realized that the Great Glen and Loch Ness could not be explained by fluvial reasoning. Sedgwick (1842) raised the objection in the case of the valleys and lakes of the English Lake District, and which he attributed to the fractures consequent upon updoming of the region. To these valid objections there was no suitable retort.

The fluvial principle of accordant river junctions, often known as 'Playfair's law of accordant junctions', although he merely restated it from Hutton, was another target of criticism. Field evidence from the Alps, an area often more familiar to London critics than Highland Britain, easily demonstrated that this was not a universal law. Playfair would have been aware of this after his continental journey, but no written reactions remain. Usually the result of differential glacial erosion at valley junctions, this objection too had no real answer until the advent of a comprehensive glacial theory.

Two other objections that were raised were two sides of the same problem: the question of massive valley infills, now termed paraglacial sedimentation, and its complement, streams that were grossly undersized for their valleys; underfit streams. These obstacles could be overcome, as they are now, by appealing to massive increases in the formative discharges for the valleys, and their infills. This argument, however, strikes at the heart of the law of uniformity as enunciated by Lyell, unless it is modified by an appeal to mild catastrophism of the type discussed by Lyell. Thus, answering the objections in this manner caused almost as many problems as the objections themselves. Significantly, however, it did slowly move the parties towards a compromise position, although this was a philosophical shift rather than the solution of specific examples.

Two other related objections centred on the existence of water gaps, and the coincidence of valleys with major structural lineaments. Hutton had discussed the true origin of the Potomac water gap, and the Wealden water gaps were homespun examples, but few geologists in 1830 could visualize the uniform operation of subaerial processes over sufficiently vast timescales with the intuitive grasp needed to appreciate Hutton's insight. Lyell attributed Wealden topography to marine processes operating on a fractured, upraised, dome.

Similarly, the influence of structure on valley trends is an idea that needs very careful treatment, and coincidence in alignment is not, in itself, conclusive evidence of causal connection. Since uniformity and fluvialism were still being disputed, the notion of adjustment to structure (subsequent streams) naturally did not occur.

Both of these objections, therefore, could be classed generally as failures of the imagination, rather than as flaws in the fluvial hypothesis. Nevertheless, their successful removal depended no less on the accumulation of field evidence than the previous four.

However, it would be a mistake to think that the average geologist, say between 1830 and 1840, belonged to one of only two camps. In fact every possible shade of opinion existed. Bakewell (1838) listed five theories for the formation of valleys: (1) unequal original deposition, (2) excavation by the rivers presently in them (Hutton/Playfair), (3) differential elevation or subsidence of the earth's crust, (4) excavation caused by the sudden retreat of the sea from the present continent, (5) excavation caused by sudden deluges that have swept over the surface of different parts of the earth. To this list can be added Lyell's Wealden hypothesis, a slow version of (4), and the strongly structural notions that Whewell and Hopkins espoused, based on mathematical theories of structural elevation, although these might be accommodated under (3).

The case of Niagara Falls forms a convenient test case for most of these views. It was discussed at length on both sides of the Atlantic with regrettably inconclusive results!

Niagara Falls: the idea of a history, the history of an idea

Niagara Falls was first seen by a white man, Father Hennepin, in December 1678, although it had been known by report for years before that. By the middle of the eighteenth century it lay on a well-travelled route to the upper lakes. Figure 4 (overleaf), from Bakewell Jr (1830), illustrates the physical geology of the Niagara River. The first informed published comment was by Captain Pouchot (1760, Eng. trs. 1781) who identified the old river banks in the environs of the Falls, implied retreat of the Falls to create the gorge, and identified the Niagara Escarpment as an old shore of Lake Ontario. The first published account in English was an anonymous letter to the 'Antigua Gazette' (1768) in which it is stated that the falls 'broke up by small degrees, to their present situation, which is seven miles higher.' This is almost word for word the view of Pouchot. These opinions probably derived from local sources for Robert McCauslin, who lived by the Falls for nine years, recorded (1793) that:

It is universally believed that the cataract was originally at this ridge, and that it has by degrees worn away and broke down the rock for a space of these six or seven miles. Some have supposed that from these appearances conjectures might be formed of the age of this part of the world.

He went on to throw doubt on this; he had noted little recession in nine years himself, and reversed the argument to show that with the earth's age fixed at 5700 years, the retreat would have had to be 66.5" a year, which he disputed was the case while he knew the Falls. McCauslin's remark probably refers to William McClay (1790) who took a local estimate for retreat, of 20 feet in 30 years, and estimated the age of the earth as 55,440 years, or, at least, as the time since the Falls 'began to fall over the ledge of rock.'

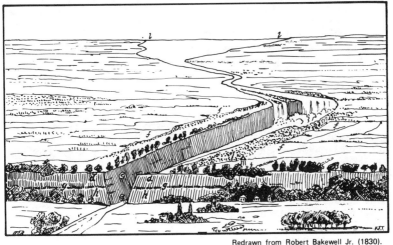

Redrawn from Robert Bakewell Jr. (1830).

LEGEND

c	Channel in hard rock	g	Loose shale	f	Alluvial or diluvial cliffs
d,h	Once united hard rock	l	Lake Erie		

Figure 4. *Bakewell's bird's eye view of the Niagara Gorge. Versions of this diagram have descended through Lyell's "Principles of Geology" to modern textbooks.*

We see in these remarks some elements of the denudation dilemma, and it is clear that local opinion, impelled by the 'obvious' physical geography of the situation, deduced what had happened and, in its untutored and ignorant way, was years ahead of academic opinion! Some of these reports came to the attention of Playfair (1802, #322) who drew the general inference that:

the height and asperity of every waterfall are continually diminishing; that innumerable cataracts are
entirely obliterated; that those which remain are
verging towards the same end, and that the Falls of
Montmorenci and Niagara must ultimately disappear.

However, the case was not as simple as it looked, for,
as the objection was well stated by Duncan (1823):

if we grant, that there was a time when the water from
Lake Erie first made a breach in Queenston Heights,
these theorists cannot refuse, that there must have been
a previous time when no breach existed. If so, where
then was the the outlet to Lake Erie? By what channel
did the waters of the great chain of western lakes,
above Ontario, find a passage to the Ocean? If these
lakes did not then exist, and if they and their outlet
were the simultaneous result of some great terraqueous
avulsion, may not it be reasonably concluded that the
whole channel of the Niagara, from the present falls to
Queenstown, was ploughed out by the same revolutionizing
struggle? that in place of being the work of thousands
of years, it may have been the work of a month, or
perhaps a day?

The objection was a telling one; clearly there could be no
smooth transition from pre to post-initiation conditions and
the lake basins presented the limnological objection
mentioned above. In addition there was the issue of how the
escarpment itself was formed.

The next phase in the interpretation of Niagara Falls
was ushered in by the camera lucida drawings made by Captain
Basil Hall (1829), and the first comprehensive geomorphological description of the gorge given by Robert Bakewell
(jr). Hall's drawings eliminated the distortions present in
artistic representations, and Bakewell's bird's eye view
(Figure 4) revealed the regional setting. His diagram was
eventually adapted by Lyell for the 'Principles' and it has
appeared in various versions in many textbooks since.
Bakewell's paper was forwarded to its publisher by his father
with a covering letter explaining that 'the subject possesses
peculiar interest at the present time, from its connection
with the enquiry actively going on in this country,
respecting the extent of atmospheric agency, and that of
rivers and torrents, in modifying the surface of the globe.'
Bakewell's paper (1830) is a minor classic. He describes the
gorge in meticuluous detail, missing only the buried gorge,
and he showed that it could not be the result of an
earthquake. Lyell used the accounts of Basil Hall and
Bakewell to prepare the section on Niagara in the
'Principles.' He used Bakewell's estimate, taken from local

inhabitants, that the recession rate was 50 yards in 40 years, and tentatively computed with Bakewell the age of the gorge as 'nearly ten thousand years' (1830-1-i-181). He added the comment that the Falls would reach Lake Erie in about thirty thousand years, but he was not averse to contemplating the catastrophic effects that the release of the waters might have on the valley below.

Lyell's authoritative publicity of Niagara Falls led to several papers in response. Coneybeare (1831), as part of a discussion on whether cataracts were relevant to the speculations of the fluvialists, wondered why the Niagara River had achieved so little compared to the River Thames, in what he imagined was the identical timespan. He was led to think that 'some partial degradation of the strata has here been mistaken for the general retrogradation.' Almost all the authors who wrote on Niagara doubted the rate of recession, even Featherstonehaugh (1831), who was well acquainted with the Falls and did believe that the gorge was due to recession. Lyell responded to these criticisms by making minor textual emendations in the 4th edition of the 'Principles' (1835), although they amounted to no real change in opinion, and he added a reference to Featherstonehaugh's paper in deference to a critic with local field knowledge.

In contrast, Fairholme (1834), writing from Britain, revealed a typically scriptural approach: a poor grasp of the physical geology was used to justify a theoretical speculation that the Falls had retreated in half the time allowed by Bakewell and Lyell! This could then be equated to a universal deluge and he finished with an invocation to De Luc, Dolomieu and Cuvier, all out-moded catastrophists, years behind the more enlightened breed in Britain who had moved on to multiple deluges.

Fairholme's paper triggered an equally ill-considered attempt by Rogers (1835), a well-known American structural geologist, to show that the gorge was merely a ravine excavated by a diluvial current. He doubted that the Falls had ever been at Queenston, but did suggest comparative work with other New York state waterfalls. In an editorial note, Benjamin Silliman remarked that a quite moderate earthquake would suffice to create the necessary gorge!

Diluvial ghosts were finally laid to rest when Lyell made two visits to the Falls during his first American visit. As with the Scandinavian uplift, he was anxious to see the field evidence and his first visit was made in the company of James Hall, the New York State geologist, and together they eliminated the idea that the escarpment was a fault. Hall pointed out the possibility of the buried gorge intersecting the present gorge at the Whirlpool, and Lyell, in a second visit lasting a week, established all the essential facts pertaining to the Falls and its gorge. He convinced himself that the Falls had eroded the gorge, but a cautious approach caused him to modify the recession rate to one foot a year, and he revised his estimate of its age to 35,000 years. Both Lyell (1842, 1843) and Hall (1842) wrote accounts of the gorge and Lyell attributed the initiation of the gorge to the withdrawal of a glacial submergence, and the escarpment to marine erosion. Hall was rather more circumspect in that direction and ordered a trigonometrical survey of the Falls to serve as a benchmark from which to establish the recession rate, and he made comparisons with other local waterfalls.

Lyell's account remained, fossilized, in all subsequent editions of the 'Principles.' As with the Weald, Lyell merely established a special case, and did not seek to generalize its fluvial conclusions. Hall has been overlooked in discussions of American fluvialism, yet his famous report on the 'Geology of the Fourth District of New York' (1843) contains a lengthy discussion of the modern action of rivers, including Niagara, illustrated by the documented retreat of waterfalls and the operation of subaerial processes. He went on to become the primary paleontological figure in North America. No significant challenge emerged to the interpretations of Lyell and Hall, and by 1845 the fluvial question was in abeyance in favour of the glacial problem.

Figure 5. *The lower falls at Portage (N.Y.) on the Genesse River from a sketch by Mrs. Hall, from Hall (1843). For details of changes to 1843 see Figure 6. Now (1983) the small prominent fall has retreated upstream over 100 metres.*

Niagara thundered on. The precise surveys initiated by Hall became a tool for the quantification of landscape in the hands of G.K.Gilbert, half a century later. A prime piece of evidence in the case for the prosecution of catastrophism was now left stranded on the brink of a breakthrough like Goat Island itself: once central to a powerful but divided stream, now shouldered to one side by the implacable recession of the mainstream - a landform acting as its own metaphor.

Fluvialism in the decades after 1840

On both sides of the Atlantic the white heat of battle abated after 1840; cooled by rival versions of the glacial theory: a smothering ice cap, submergence in a glacial sea, or both. However, James Hall stimulated by Lyell's visit prepared a thorough account of the modern action of rivers for his 1843 report. He specifically addressed the question of 'whether the present streams can have formed the channels in which they now flow.' His answer was positive for he noted specific changes, in living memory, at the lower falls of the Genesee River at Portage, and remarked that:

A,B.   Represents the width of the chasm at the top.

a,a.   The platform or bed of the stream, over which the water was originally precipitated ninety-six feet to the level of the river below the falls. This platform *a a* was formerly continuous to *a'*.

b.   The narrow channel of recent excavation.

d,d, and c. A recent gorge, separating the small island from the main bank.

Figure 6. *Hall's diagram to explain changes up to 1843 on Figure 5.*

Our theories of modern operations make provision for wide sweeping deluges, for immense excavating waves, for hemispheres of ice, the upheaval of mountain chains, and the transportation of ice floes from the frozen islands in our own latitudes; but we have almost forgotten the quiet operation of running streams, and the freezing of water in fissures of hardened rocks (Hall 1843, p365).

Regrettably, Hall's work had no apparent impact, perhaps because there was no follow up, and this aspect of his work has been neglected by historians of geology. Within a few years another independently minded geologist, James Dana (1849), from observations made in Australia and Hawaii during a tour of duty with the Wilkes Exloring Expedition, began preaching unadulterated fluvialism. Chorley et al. (1964) note that although his reputation grew slowly in the United States, he was virtually unknown in Europe until the issue had been resolved on other evidence, some of it American.

In Britain, a number of geologists wrote about the efficacy of meteoric agencies: Prestwich, de la Beche, Godwin Austen, Whitaker, Jukes and Greenwood. Of these, Greenwood has often been given pride of place in modern writings for his stand on behalf of pure fluvialism (Stoddard 1960, Chorley et al. 1964). However, Davies (1969) has argued persuasively that Greenwood's naive writings had little real effect in the period before 1862 and that he was largely ignored. Certainly his actions, typical of the over-enthusiastic amateur, were those of an agent provocateur. They remind one of the visitor abroad who attempts communication by shouting louder and longer, in his own language. Greenwood's views, first publicized in 'The Tree Lifter' (1853), but better known through 'Rain and Rivers, or Hutton and Playfair against Lyell and all comers' (1857) scarcely went beyond reiterating the logical consequences of rainfall in the presence of gravity. But the real issue was what precise landforms this effect could create over aeons of time. Few doubted the day by day behaviour of rain and rivers in carrying sediment to the ocean, and quite refined estimates of erosion rates had been made by mid-century. What was required was a much vaster conceptual scheme, which would show how huge portions of the surface would develop in time; in particular the explanation of topographical anomalies in river systems required more than simple invective and a mindless faith.

Far more significant, and a much more positive step on behalf of fluvialism, was the paper by Jukes (1862) on the rivers of southern Ireland. Yet, Jukes was merely reiterating for rivers in the south of Ireland, where transverse gorges cut across the clear lines of geological structure, expressed as longitudinal ranges of hills, the transatlantic truth discerned by Hutton, seventy years before, for the Potomac in its water gap through the Appalachians. Jukes made the additional observation that the original river pattern of the superimposed river system, developing on an initial surface of low relief created by marine denudation, could be reconstructed by mapping the alignments of river gaps and their initial headwaters, even though subsequent development along the strike vales has altered our perception as to which is the master stream. Furthermore, as a professional geologist, he was able to show that 'these ravines are not fractures caused by internal disturbances;' moreover the transverse ravines 'are often tortuous, precisely like the bed of a river worn by its own action into the rock below.'

Jukes's insight proved appealing to his professional colleagues in the Geological Survey; Archibald Geikie who specifically acknowledged Jukes' insight in working out river histories (Geikie 1865), and particularly Andrew Ramsay who had long favoured the marine erosion of upland plains in Britain, the starting point of Jukes' analysis. Ramsay (1862) provided, in his turn, the crucial paper on the glacial erosion of bedrock basins which freed geomorphology from the limnological objection. Within a single year, therefore, fluvialism was re-established on a base of careful fieldwork, and glaciers had significant, and selective, erosion added to their repertoire, in addition to their undoubted depositional capabilities. This recognition obviated the necessity to call upon grounded icebergs in a glacial sea as the cause of glacial striations. Deluges, single or multiple, were banished, and Lyell's 'Principles' in its later editions began to look dated, if not antiquated. Many of its incidental facts remained correct, even neglected, but its conceptual framework for regional landform development was rapidly becoming outmoded. In particular it lacked any scheme whereby the evolution of an abitrary piece of landscape might be worked out, and it relied too heavily on a simplistic Huttonian notion that different structural units could behave in tectonically independent ways with little thought as to how this behaviour might correlate over space.

In the years following Jukes's paper, fluvialism was widely adopted in Britain, helped by the Geikie's 'Scenery of Scotland' (1865) and Ramsay's 'Physical Geology and Geography of Great Britain' (1863), both of which were firmly fluvial in outlook. The scenery of the Weald was given a fluvial interpretation by Foster and Topley (1865) and in 1868 Geikie published a famous paper in which he showed from a quantitative analysis of erosion rates and river loads that:

> before the sea could pare off more than a mere marginal strip of land 70 to 80 miles in breadth, the whole land would be washed into the ocean by atmospheric denudation.

A little later, Huxley's 'Physiography' (1877) took the Thames Basin as the basis for the study of landscape under the influence of atmospheric agencies (Stoddart 1975). The continent saw a resurgence of fluvial interest from independent sources. Rutimeyer (1869) in Germany sought to explain Alpine valleys entirely by fluvial processes. His advocacy of this theme was so strong that, together with the alpine geologist Heim, he was responsible for providing the impetus for a small but vocal school of glacial 'protectionists' whose British adherents included Garwood and Bonney. De la Noe and De Margerie (1888) in France, although they acknowledged Ramsay, made it clear that their primary sources lay in the United States, where Dana's 'Manual of Geology' (1863) contained discussions of river long profiles based on Du Buat, and his own tropical field experience, in Hawaii and Australia.

A new perspective often requires new evidence and in the earth sciences this usually implies a shift to new terrains. The Old World was to give way to the New and the exploration of the American west after 1860 provided the stimulus that led to the development, from new and unambiguous landscapes, of novel and comprehensive concepts about the evolution of regional landscapes.

BEN SLIOCH—A MOUNTAIN OF TORRIDON SANDSTONE.

Chapter Six

CATASTROPHES IN METAMORPHOSIS: FLOODS, FIRES AND SHAKINGS

> *Others have referred to the Deluge, − a convenient agent in*
> *which they find a simple solution to every difficult problem*
> *exhibited by alluvial phenomena.*
>
> *(Charles Lyell, 1833, 1-III-148)*

The attribution of landforms to the effect of floods, however
these are construed, is not surprising on a planet whose
surface is seventy percent water. The idea harks back to the
earliest Greek literature, but the Mosaic account in the
Bible ensured an entrenched position for 'Floods' in the
intellectual development of western literature. Woodward
(1695) made typical, and essentially magical, use of the
Mosaic Flood in his 'Essay towards a Natural History of the
Earth,' and he was careful to forwarn his reader:

> It will perhaps at first sight seem very strange, and
> almost shock an ordinary Reader to find me asserting, as
> I do, that the whole Terrestial Globe was taken all to
> pieces and dissolved at the Deluge, and Particles of
> Stone, Marble, and all other solid Fossils dissevered,
> taken up in the Water, and there sustained together with
> Sea-shells and other Animal and Vegetable Bodies: and
> that the present Earth consists, and was formed out of
> that promiscuous Mass of Sand, Earth and Shells, and the
> rest, falling down again, and subsiding from the Water.

There was little that rational argument from physical
principles could do against such confident assertions, and
yet in 1838 Bakewell was protesting, only a little late in
the day, against those who would remove denudations by 'the
vague supposition of diluvial currents.'
The seventeenth century, with the few honourable
exceptions already encountered, assumed as an article of
faith that the Flood was responsible for topography. Not the
slightest attempt was made to apply hydraulic analysis to its
supposed effects and it was only the increasingly detailed
field evidence in the first half of the nineteenth century
that caused a shift in emphasis towards smaller, more
localized, deluges. Woodward, it is true, had separated the
Mosaic Flood, which he insisted was universal, from the
smaller historical floods in Greek literature, such as the

Deucalion flood. However, the slow transformation, and eventual abandonment, of the the Mosaic Flood was forced upon
geologists by field evidence.

The catastrophic twin of flood is fire, but this was
viewed as a less universal agent in the creation of landforms
than water. Although the solidification of lava at volcanos
was not denied, its identification with basalt in extinct
volcanic regions, or in exposed rock sections, was vigorously
disputed by the Wernerian school of geologists. There were
again honourable exceptions: the Frenchmen Guettard and
Desmarest, and the Scotsmen Hutton, Hall and Playfair.
Consequently, while water was over-emphasized, fire was
under-played. Earthquakes attracted more attention, as is
understandable. Their propensity to destroy habitation has
made them a focus of attention throughout history: Pliny
devoted much space to them, Hooke (1705) wrote his
'Discourses' upon them, and fired by the Lisbon earthquake in
1755, Michell (1760) wrote an influential essay linking their
operation to the generation of steam when atmospheric waters
encountered fiery regions within the crust.

Because the power of earthquakes to disrupt the surface
was beyond dispute, and as there was no way of predicting the
specific power a particular earthquake might be supposed to
have in any region, this agent, too, took on quasi-magical
powers. Earthquakes could be called upon to account for any
real or imagined feature of surface topography. In the
nineteenth century, appeals to their power became more
refined: in the hands of Lyell it became apparent that
repeated applications of small earthquakes could be just as
effective in moving the crust up and down, as the invocation
of an entirely imaginary one of gargantuan proportions.

In the previous chapter I showed that the evidence of
volcanic regions, in the skilled hands of Lyell, dispelled
assertions of a huge and recent Mosaic Flood. Likewise he
tamed earthquakes by applying uniformity. In the remainder of
this chapter I would like to demonstrate in the context of
actual field problems how various workers grappled with
catastrophism, mainly in the guise of floods, since these
were by far the most ubiquitous agents. First, however, it is
important to re-establish in the reader's mind the fact that
in the field there was compelling evidence for catastrophic
events. The recognition that this evidence constituted a
problem which the simplistic application of the principle of
uniformity could not dispel, forces us to acknowledge the
intuitive powers of understanding possessed by many of the
early geologists.

'Terreins de transport'

Because the development of geology took place in high
latitudes, and almost all these areas were affected by

Quaternary glaciation, that deposit now loosely known as 'drift,' was well-known to field geologists and posed a puzzling conundrum. Typically, the underlying solid rocks stopped short of the surface, to be replaced by an endlessly varying deposit composed of anything from pure clay to coarse boulders, with every possible gradation between.

Chapter Three mentioned Hutton's attempt to trace the quartz-rich gravel spreads of the English Midlands to their source, and in his 'Theory of the Earth' (1795) he discussed at some length the particular problem of large erratic blocks. Playfair (1802, #366), in notes on the 'Transportation of stones,' specified a sedimentary assemblage which would argue against a fluvial origin:

> Lastly, if there were any where a hill, or any large mass composed of broken and shapeless stones, thrown together like rubbish, and neither worked into gravel nor disposed with any regularity, we must ascribe it to some other cause than the ordinary detritus and wasting of the land. This, however, has never yet occurred; and it seems best to wait till the phenomena is observed, before we seek for the explanation of it.

He went on to explain that he was perhaps tilting at windmills since 'neither Pallas, nor Saussure, nor Dolomieu, nor any other author who has espoused the hypothesis of such causes, has explained his notions with any precision.'

Playfair's description would presumably exclude a great many bedded paraglacial sediments, but even so we may suspect that he was deficient in serious field work in glaciated terrain since his description is really a very good one for many glacial tills. It was partly to dispel the semantic overtones that most descriptions of these superficial deposits possessed, that Bakewell (1838) yearned to use the French term 'Terreins de transport.' Instead, Murchison's use of the term 'Drift' (1839) became general, and still survives in the Drift sheets of the Geological Survey.

The sudden transition from unconsolidated, sometimes fossiliferous, and often chaotically bedded sediments to the orderly arrangements characteristic of the underlying 'solid' geology did pose a problem, and it was not easily resolved by adherents of fluvialism. It was precisely because of its chaotic appearance that the Flood, or floods, seemed an obvious recourse: what was needed was an agent of considerable power and relatively instantaneous action.

JAMES HALL OF DUNGLASS (1761-1832): A THEORY OF A DELUGE

James Hall's adoption of Huttonian theory was nothing if not considered. He himself recorded (1805) that it took place 'after three years of almost daily warfare.' Yet his conversion never did include an acceptance of what he termed 'diurnal causes' as significant in the creation of major landforms. Playfair, in the paragraph after the one quoted above, indicated that he expected a broadside from Sir James on the issue of landforms in the Edinburgh region and this was duly delivered in 1812, and published in 1815. He differed from Hutton on 'the particular mode by which he conceived our continents to have risen from the sea.' He believed it was in a series of spasms, not 'by a motion gentle, as to leave no trace of the event, and so as to have had no share in producing the present state of the earth's surface.'

To modern eyes Hall's paper is a masterpiece, not for its diluvian conclusions, but for his detailed and sensitive study of the topography around Edinburgh, and for the way in which it is integrated with prevailing theory, and is related to more general observations, and laboratory experimentation. The mechanism of a diluvial torrent, as Playfair had asserted, had never been made clear. Hall rectified this by developing a theory of waves of translation caused by the sudden injection of material into the sea bottom. He supported this theory with observations made of the tsunamis associated with the Lisbon earthquake in 1755, and by laboratory experiments in which he exploded charges of gunpowder beneath standing water! Regrettably he had no means of recording such events and his experimentation had less impact on geomorphology than his petrological investigations had for geology. Hall was an experimentalist rather than a theoretical physicist, and on the basis of his intuition he believed that:

> It would be impossible for water of any depth whatever, or moving with any velocity, to carry blocks of such magnitude to such situations; and the consideration is of so great an importance, that I am induced, in attempting to unite the ideas of SAUSSURE with those of HUTTON, to retain part of the system proposed by M. WREDE, in so far as to consider the granitic blocks as having been made to float, by means of a mass of ice attached to each (Hall 1815, p156).

It is unfortunate that Hall did not avail himself of the continental hydrodynamical literature so abundantly available in Playfair's library, since a straightforward application of physics and hydraulics (along the lines of Graf, 1979a) would have shown that even the huge Jura erratics could, potentially, be moved by a deep enough flow. However, he did consider the general pattern of deposition around obstacles in fluid flow and applied these ideas to the streamlined landforms in the area around Corstorphine Hill in Edinburgh.

He noted, astutely, that diurnal causes were acting down the present slopes, not along them. In addition:

> It is in vain that a vast duration is ascribed to the influence of an agent, unless it can be shewn, that its action has a tendency to produce the alleged result. If it has a tendency to produce a different result, that difference would be augmented in proportion to the duration of the action. Now, the diurnal operations are everywhere found in the act of corroding and altering the forms here alluded to (as we see on the road to Dalkeith, just as we leave Edinburgh); but they are nowhere seen to produce them (Hall 1815, p178).

Hall then noted the relationship between meander wavelength and stream size and compared the Water of Leith, 'a paltry brook,' with the straightness of the streamlined ridges whose radius 'mathematically speaking .. is of infinite length' and he opined that a stream had flowed over the area 'disregarding objects by which the Nile or the Ganges would have been turned out of their course.' Streamlined features in the neighbourhood of Noblehouse he termed 'craigs and tails,' a terminology still used in glacial geomorphology, and he drew attention to the fact that some minor features, especially the striations he observed on 'dressed surfaces,' showed a greater sensitivity to surface slope than larger scale forms and argued that this reflected a less powerful stage in the flood. With remarkable intuition he deduced that the flood must have been at least twice the height of Corstorphine Hill or 'it must have locally obeyed the inclination of these surfaces,' and so:

> if we consider our stream as .. one thousand feet, we shall probably not go beyond the truth. This height is nearly sixteen times the altitude of the wave at Cadiz, which was sixty feet. The phenomena of the Alps, though imperfectly known, indicate a magnitude double of this, by the mere position of the blocks on Jura, two thousand feet above the level of Lake Geneva. While I thus, however, ascribe great magnitude to the diluvian torrent, I am on my guard against excess of such impressions, and the means have already occurred, as I shall presently state, by which limits are assigned even to this colossal agent (Hall 1815, p197).

In nearby areas Hall discovered that on a larger scale still the torrent had been deflected by the general pattern of the Firth of Forth estuary, and so it was apparent, as he had suspected, that the torrent was not limitless in size.

The field evidence used by Hall was tabulated and keyed into a base map at a scale of a little over 3 inches to the mile. The data he gave is reproduced as Figure 7 and probably represents the first published geomorphological data. A detailed description is given in his text of each site.

SPECIMENS.

No. 1. Craigleith Quarry, from   -   W. 5^0 S. to E. 5^0 N.

  2. Maiden Craig,   -   -   W. 5^0 S.

  3. Ravelstone old farm-yard,   -   W. due.

  4. Well Craig, near Craigcrook,   W. 20^0 S.

  5. Craighouse Quarry,   -   W. 5^0 S.

  6. North of ruin at Dean of North
       Clermiston,   -   -   W. 10^0 S.

  7. North-east of cottages there,   W. 10^0 S.

  8. Middle of the North Hill Park,   W. 10^0 S.

  9. South side of ditto,   -   W. 10^0 S.

 10. Bare space west of summit,   W. 10^0 S.

 11. Summit of the hill,   -   W. 10^0 S.

 12. South-east corner of South Mid-
       Hill Park,   -   W. 15^0 S.

 13. Summer-house on second sum-
       mit, belonging to Ravelstone,   W. 15^0 S.

 14. Sheep Park of Corstorphine
       Hill House,   -   W. 8^0 S.

 15. Below Murrayfield Quarry, east
       of Belmont,   -   W. 15^0 S.

 16. Dickson's Craig, Barnbugle,   W. 3^0 S.

 17. Redhall,   -   -   W. 8^0 S.

 18. Ravelrig,   -   -   W. 15^0 S.

 19. Kingston, near North Berwick,   W. 15^0 S.

Figure 7. *The table of data for field localities, west of Edinburgh, which illustrated streamlined, oriented landforms and "dressed" and "scratched" rock surfaces. Localities are discussed in detail, and keyed to a base map in Hall (1815). This is probably the first published geomorphological field data.*

We now recognize his evidence as typical of glaciation, and with the subsitution of 'glacial' for 'diluvial' the paper makes perfect sense today. His attribution of the surface 'dressings' (polished glacial pavements) to the combined action of water and sediment, might still find favour today, in a subglacial context. The deduction of diluvial depth is uncannily accurate in the context of ice thickness, and the whole paper still makes extraordinary reading. In conclusion, Hall suggested the application of the principles he had deduced as a 'means of ascertaining the direction of diluvian inundations across the great continents. By a comparison of directions, these tremendous agents can be traced to their source.' Thus, with characteristic frankness, Hall was suggesting a regional analysis that would help to elucidate problems posed by a local study. But nobody took up his challenge, and four years later Buckland, in his inaugural lecture as Professor of Geology at the University of Oxford, resolved in his own mind the conflicting field evidence with a single Mosaic Deluge (Buckland 1820). He reinforced his position four years later with his 'Reliquiae Diluvianae' (Buckland 1823) but the increasing weight of field evidence forced the confrontation outlined in Chapter Five between Lyell and Murchison, and Buckland and his assorted diluvian disciples.

MULTIPLE DELUGES, VALLEY EXCAVATION AND MARINE SUBMERGENCE

Buckland's single Deluge position was upheld on the continent by Cuvier, with intellectual support from Elie de Beaumont who considered that mountain chains were formed in short-lived catastrophic upheavals: hence the uplift of the Andes might have caused the great Deluge. However, evidence from the Auvergne volcanic region was enough to force the English catastrophists into retreat, and Figure 8 shows, in stark simplicity, that there had been repeated phases of valley cutting, gravel deposition and burial by lava flows.

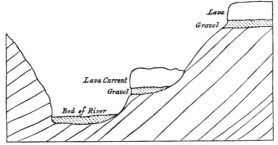

Lavas of Auvergne resting on alluviums of different ages.

Figure 8. *The evidence from Auvergne which forced Diluvialists to shift to a multiple flood position (Lyell 1833-1-iii-267).*

111

The catastrophists were forced to accept multiple deluges, and in the light of the discussions about uniformity, and the origin of valleys, their arguments taken as a whole underwent a significant metamorphosis. In the Lyellian scheme, the upheaval of structural domes such as the Weald and the Lake District, naturally led to suggestions that valley lines owed much to fractures, fissures and dislocations, aided by waves of translation, a notion revived with impressive mathematical apparatus by Hopkins (1842). The removal of detritus and solid rock, 'denudations,' might also take place beneath the sea due, it was thought, to the action of tides and currents (Bakewell 1838, Murchison 1839). Alternatively, slow emergence of a landmass from beneath the ocean enabled the efficacious processes of the coast to effect the erosion of escarpments, and the removal of river borne sediment (Lyell 1833-1-iii-285).

Thus, within the overall framework of Lyellian theory, the ingenious thinker found it possible to argue for that particular combination of forces thought appropriate to the problem. Some measure of theoretical unity was thereby achieved, even though this was less apparent to Lyell's contemporaries than it is to us now. In this connection I will examine Murchison's response to the field evidence of the Welsh borderlands, Siluria as he termed it, in the light of his work with Lyell in the Auvergne, and the fact that he remained a lifetime catastrophist and neo-diluvialist.

Murchison's 'Silurian System' - local catastrophism

Murchison remained proud that he and Lyell had banished the idea of a single universal, Mosaic Deluge (Murchison 1839, footnote p522). The last section of his 'Silurian System' deals as carefully with the surficial deposits as with the Silurian rocks. He proposed to replace the term 'diluvium' with 'drift' to avoid the connotations of the former term. Careful field work enabled him to distinguish two local drifts, and a 'northern' drift. Within the latter he drew attention to the many, and large, erratic blocks. He showed that the drifts showed no relation to the present drainage pattern, and he saw an analogy between Siluria and the Auvergne in many respects concerning their drift deposits. His experience with sedimentary characteristics in the older rocks, and their relation to depositional environments ensured that he was responsive to such evidence in the drift and so, of a finely comminuted and well-bedded sand, he noted 'we cannot but attribute their deposit to long-continued periods of sub-aqueous accumulation, and not to any sudden and transitory rush of waters.' This he distinguished from hillocks and heaps that are 'not arranged in terraces, as if produced by any possible fluviatile action, but are placed high up in confused and irregular heaps in the mountain coombs.' These deposits, and the landforms they occupied, he believed to be due to the action of currents and tides beneath the ocean surface. It was typical of such claims that no clear distinction was made between the erosional and depositional characteristics of the supposed waters.

The erratic blocks embedded in the northern drift posed an additional problem, and to solve it Murchison was happy to rely on the notion that the blocks had been rafted by a skirt of ice. For this idea he acknowledged an impressive list of authorities going back through Lyell to Sir James Hall and Professor Wrede (of Berlin) whom Hall had quoted. Wrede's idea probably came from experience on the Baltic shore, and Murchison also saw fit to publish, in an appendix, Captain Bayfield's letter to Lyell detailing the ice rafting of boulders in the St Lawrence River. Lyell subsequently gave this prominence, and a specially annotated plate, in the 'Principles.' Lyell's trip to North America in 1841/2 further confirmed the view that erratic blocks were ice rafted, together with a bevy of reports of boulders observed on icebergs, in northern and southern seas.

Murchison's explanation of the local source of the drifts depended on adopting a version of Leopold von Buch's hypothesis of 'craters of elevation.' Murchison identified local structural domes, such the Woolhope dome, whose 'cratiform' topography was thought to reveal sudden uplift. The immediate area was therefore swept clear of 'drift' which accumulated in the surrounding areas. Murchison also thought that volcanic outbursts could produce similar effects, and he envisaged enormous changes in the relative level of the land in association with these tectonic movements. The finding of marine shells in drift on the flanks of Moel Tryfan to a height of 1700 feet (Trimmer 1832) helped confirm this view.

From this review it is clear that Murchison envisioned an extremely dynamic drift environment, intimately connected with structural features in the underlying Silurian. He was adamant that the modern action of 'the feeble Silurian rivers' and other subaerial agencies was, in comparison, of little consequence in the creation of the surface features. He considered, and dismissed as 'peculiar opinions,' the ideas of continental glaciation beginning to circulate in the late 1830's. He never relinquished his catastrophic views, and he complained in later years that his delimitation of the Russian drift had been neglected by later workers. The idea of a marine submergence was, however, purely Lyellian and it had years of useful life left in it before it ceded right of way to the notion of land-ice glaciation.

A fascinating transatlantic reflection of Murchison's 'Silurian System' can be found in the famous 'Geology of New-York' authored by James Hall (1843). This James Hall was no relation to Sir James Hall of Dunglass, and as I have already noted his geomorphological work has been overlooked.

James Hall of Albany: The 'Silurian System' reflected

James Hall (1811-98) was assigned the western section of New York State when he was inexperienced, and because the geology was thought to be 'easy.' As a model for his report on a predominantly Silurian terrain it was natural that he would use the monumental, and recent, report on the identical rocks in England and Wales: Murchison's 'Silurian System' (1839).

An additional influence on Hall was the visit of Lyell in 1841 when they examined the Niagara Gorge together. By this time Lyell had embraced, and then rejected, the full glacial theory. Instead he promoted the notion of limited mountain glaciation, and a widespread marine submergence replete with boulder-laden icebergs. Although Hall modelled the structure of his report in some detail on Murchison's book, he began to have difficulty with his theoretical views:

> I may here remark, that these investigations were commenced with a belief in certain views and theories regarding the production of the drift; but as my observations progressed, the difficulty of reconciling the facts with these pre-conceived notions became constantly more apparent (Hall 1843, p324).

Despite limited time, and an inability to revisit critical localities, he succeeded in throwing some serious doubts on a number of problems; for example, the precise mechanisms by which boulders in icebergs might be supposed to produce striations, and how it was that the valleys of the Finger Lakes region maintained such an even width for such a distance if they had been excavated by submarine currents, or by the waters of a retiring ocean. Unfamiliarity with Agassiz's glacial theory led him to reject it since he thought, like Lyell, that the creation of the ice sheet depended on the presence of mountains, and that glacial ice merely occupied existing valleys and could not erode them.

Pre-eminent in Hall's report is the quality of his field observations. He was careful to separate weathered residual boulders which he had seen in Virginia and North-Carolina from genuine erratics and his attention to detail is evident in the following quotation:

> I attach very little importance to the supposition that boulders of granite have been worn smooth and striated upon one side while fixed in a floating mass of ice, and in that way worn down while rubbing over a stony bottom. Some boulders of this kind, which I have seen, are less than a foot in thickness, and two feet in length. Now is it possible that such a boulder, having rounded edges, can be fixed in a mass of ice, so as to allow of such force been applied to it, without falling out, unless

the pressure were constant? The beds of many of the streams in the granite regions of New-York are literally paved with boulders, which remain fixed in certain positions, while any fresh accumulations of stones and earth, with ice and water, pass over them, rendering the upper sides very smooth, while the lower may be little worn. Can such occurrences offer any explanation of this apparent polishing on one side by transportation? (Hall 1843, footnote p333)

Hall called upon stream bed armouring, rather than the hypothetical mechanism of icebergs to polish rocks! Regrettably, he never pursued these interests in later years although his seminal contribution to the theory of synclines and mountain-building ultimately was to help change the diastrophic basis of geology, always a fundamental element in geomorphological theory.

Hall's guarded puzzlement over the causes of drift topography was not shared by most of his North American contemporaries. Silliman, who had heard Playfair, Hope and Murray lecture on geology in Edinburgh, was an unashamed diluvialist. In a lengthy appendix attached to the third American edition (1839) of Bakewell's 'Introduction to Geology,' he reveals himself as still attached to the ideas of Buckland from fifteen years before. The previous chapter revealed that Rogers had attributed the Niagara Gorge to a diluvial current, to which Silliman added a moderate editorial earthquake for good measure!

Lyell's first North American visit went some way towards effecting a conversion in the theoretical views of North American geologists: mainly in the direction of accepting the idea of a recent glacial submergence. Lyell was able to confirm by his own observations that rocks floated out to sea on ice floes. His theory was a compromise, apparently well supported by evidence, between simplistic diluvialism, and the extremes of Agassiz's new glacial theory. Agassiz came to settle in the United States in 1847 but it was many years before the glacial theory was fully adopted. Eventually the shift in thought in North America was towards purely fluvial processes and hypothetical marine submergences died a natural death. Along Atlantic coastlines, however, the idea underwent yet another metamorphosis.

UPLAND PLAINS OF MARINE DENUDATION

The western coasts of Britain provide ample evidence of the power of the sea, and the presence of undoubted raised beaches on most British coasts leaves little doubt that in a limited sense the island has undergone marine submergence in the recent past. These remnants were extensively documented for Britain and selected foreign localities in 'Ancient Sea Margins' (Chambers 1848). Together with the ambiguities introduced by the presence of recent marine shells in the 'northern' drift, this evidence was quite sufficient to perpetuate the submergence doctrine in British geomorphology for many more years. Lyell was an active and influential spokesman against the full glacial theory, and Murchison, who became Director General of the Geological Survey, lent considerable experience and authority to 'neo-diluvial' views. Jukes maintained exactly this position in his 'Student's Manual of Geology' (1857). Even as late as 1869, Daniel Mackintosh wrote a book, dedicated to Murchison, attributing virtually all topography in England and Wales to the action of submarine and coastal processes.

A new logic for marine submergence

The idea that marine submergence shapes landforms proved capable of considerable refinement, when approached from a more logical viewpoint than mere assertion. Andrew Ramsay (1814-91), later to become Director General, was appointed to the Geological Survey in 1841 and his report on South Wales (1846) revealed yet another metamorphosis of deluge hypotheses. He was struck by the eveness of the skylines despite considerable structural complexity in the underlying rocks. A knowledge of the morphology of raised beaches suggested to him that a slowly rising sea could gradually planate a landscape to create a broad level bench, backed by a cliff. This idea differed from Lyell's notion of landscape dissection as land emerged from the sea since it had no place for valley formation; quite the opposite, it predicted the formation of plains, subsequently uplifted by tectonic processes, to give the level surfaces truncating the complex geological structures so prevalent in Wales.

This concept was to have striking implications since the development of landscape from an initial surface of low relief was to underlie British thought on geomorphology for more than a century. It heralded the first truly radical shift in thinking since the days of Hutton and Playfair by creating the clean slate upon which the differential processes of subaerial erosion could operate to etch fluvial landscapes. In this respect it was fundamental to Jukes's paper (1862) on the rivers of southern Ireland. Ramsay, in embracing fluvialism, subsequently retreated from the full development of the marine planation theory, but its impact on thought has remained evident down to the present day in Britain where the question has always been, how deeply was Britain submerged, first by Cretaceous and then by late Pliocene seas?

A new and rational approach to landscape evolution demanded in its turn an agreed diastrophic model.

A new diastrophism and global eustasy

Until the last decades of the century geological theories of mountain-building and continental uplift were in a state of anarchic disarray. However, with catastrophes tamed, the ice age revealed (Chapter Seven), and fluvialism re-developed (Chapter Five) the time was ripe for a long overdue reconciliation between physiographic geology, geomorphology as it was called after 1890, and the newly-developing paradigm that controlled earth history. The new ideas, largely based on a lifetime's work by Suess (1831-1914) are well-expressed by Chamberlin (Chamberlin and Salisbury 1909, p539):

> Not only are inequalities necessary to the existence of the land, but these inequalities must be RENEWED FROM TIME TO TIME, or the land would soon, geologically speaking, be covered by the sea. The renewal has been made again and again in geological history by movements that have increased the inequalities in the surface of the lithosphere. With each such movement, apparently, the oceans have withdrawn completely within the basins, and the continents have stood forth broadly and relatively higher, until worn down again. This renewal of inequalities appears to have been, in its great features, a PERIODIC MOVEMENT, recurring at long intervals. In the intervening times, the sea has crept out over the lower parts of the continents, moving on steadily and slowly towards their final submersion, which would have inevitably have been attained if no interruption had checked and reversed the process. These are the great movements of the earth, and in them lies, we believe, the soul of geologic history, and the basis of its grand divisions.

Chamberlin (1909) regarded this diastrophic model as so fundamental, since it was based on a physical theory of a shrinking earth, that he proposed to use it, rather than palaeontology, as the ultimate basis of correlation. Specifically he connected 'correlation by base-levels ..', one of the triumphs of American geology' with 'correlation by its complement, transgressive deposits on a base-level' and because base-leveling filled the ocean with sediment, a slowly rising sea level 'is thus brought into active function as a base-leveling agent.' The whole process implied 'a homologous series of deposits the world over.' America geologists differed in some minor details from the proposals of Suess. In particular, absolute stability of the continents was slowly replaced with the concept of 'epeirogeny' derived from the field work of Gilbert (1890) on the crustal deformation around Lake Bonneville, where lake shrinkage led to isostatic rebound. Indeed isostasy, epeirogeny and the complications of glacial eustasy and isostasy eventually undermined the whole basis of the model.

It was Suess (1883-1908) who first pointed out the worldwide Cretaceous transgression that was terminated in North America by the Laramide orogeny, but Pirsson and Schuchert (1915) note that 'periodic adjustment in the earth-shell of North America is recorded by at least fourteen times of mountain making.' The Suess eustatic model was propagated by an extremely influential French translation by De Margerie (1897-1918) which, with much additional material became definitive, and an English translation by Sollas and Sollas (1908-1924).

Several key concepts of crucial importance to geomorphology are evident from these considerations. Orogenic uplift was a 'rapid' geological process separated by long periods of quiescence, accompanied by a slow rise of sea level, and revealed in the geological record as areally extensive unconformities. Sea level, which controlled base level, was diastrophically controlled and globally synchronous. On these premises local high relief was either the result of recent orogeny, or of local epeirogeny with no global significance.

The basic tenets of the model outlined by Chamberlin were adopted by Davis (1889) for his Geographical Cycle (Chapter Nine), although Davis always preferred local epeirogeny to global eustasy for changes of base level. The French adopted the Geographical Cycle from Davis before the First World War, and after it Baulig (1928) made the Suess eustatic model central to his conceptions of Tertiary landscape evolution in France. His London lectures (1933) on 'The Changing Sea Level' (1935) served to anchor British thinking on denudation chronology, with its emphasis on external changes of sea level, up until the early 1960s.

The fundamental premises of Suess began to be undermined toward the end of the 1930s when the widespread significance of epeirogeny became fully appreciated, together with the complementary notion of crustal isostasy with its theoretical potential for vertical motions of the continents (Chorley 1963). Its final downfall was delayed by the Second World War but by 1949 Gilluly had proposed the radical alternative that diastrophism was continuous in geological space/time.

Within a century paroxysmal Mosaic Flood waters had been transformed, almost beyond belief, into a uniformitarian theory of global eustasy.

Diagram to illustrate the formation of the river-valleys of the South-west of Ireland.

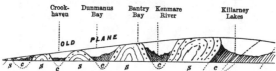

Chapter Seven

THE GREAT ICE AGE REVEALED: A SLOW CATASTROPHE?

Alluvial hills of Wayne County.
Direction N 10° E. and S 10° W.

(James Hall, "Geology of New York," 1843, Plate IX)

General acceptance of the reality of the recent Ice Age, after 1862, was one of the crowning intellectual achievements of nineteenth - century geology. The blend of factual investigation and imaginative insight that it represented was unparalleled in the history of geomorphology up that point. Neither element had been lacking before, but previously they had not been combined with such success. Almost all traces of deluges were obliterated yet an Ice Age could explain anomalous drift deposits, curiously shaped landforms, and deeply excavated rock basins with equal success. It justified those who felt that mere application of the principle of uniformity was not entirely sufficient, since it implied that in the recent geological past there had been a radically different climate, and an agent of enormous mechanical power. Yet, the recent existence of widespread land ice, over regions now lacking glaciers, was sufficiently close to existing glacial terrains in time, space, and typology, for the uniformitarians to be able to stretch their principles to include the new agency. Naturally, the idea was born and nutured in the Alps, but Great Britain became an important testbed for the glacial theory since it was an island, possessed mountains, and now lacked glaciers. In addition, British geology led the world in the first half of the nineteenth century and acceptance in Britain would give a universal stamp of approval. North America took its turn as a learning ground for glacialists once it became apparent that the Ice Age had involved ice sheets that were relatively independent of an origin in high mountains.

CONCEPTUAL PRECURSORS

The extension of the hypothesis that glaciers are active geological agents of erosion and deposition can be divided into four stages: (1) the description of existing glaciers, their modes of operation, and associated landscapes, (2) the

119

recognition that landforms within a few miles of ice margins are of glacial origin, (3) the notion that landforms tens of miles away, although in the same general region, may have a glacial origin, and (4) the extrapolation of the glacial hypothesis to areas now completely lacking active glaciers.

The history of the glacial idea roughly parallels these steps in the imagination. The period up to the end of the eighteenth century corresponds roughly to the first two stages. The first thirty or forty years of the nineteenth century would represent stage three, while the final stage was reached by about 1840. These time periods correspond approximately to the initiation of the stages as issues of public debate; the ideas may have arisen earlier but been neglected, and certainly there were considerable lags before the final stage was universally accepted.

During the seventeenth and eighteenth centuries the Alps experienced climatic deterioration and a significant expansion of the glaciers (Grove 1966); a period that has been dubbed the The Little Ice Age. In the nineteenth century the glaciers either held their ground or began a phase of retreat that has continued up to the present. Not surprisingly, the alpine peasantry knew well the behaviour of glaciers and the characteristic deposits associated with them. As early as 1723 Scheuchzer, a Swiss naturalist, proposed that the motion of glaciers was due to the expansion of meltwater when it froze to form ice in the crevasses of the glacier. A few years later Martel (1744), a Swiss engineer, described the ability of glaciers to move enormous blocks of rock, and to leave them beyond the present limits of the ice. Later in the century Saussure (1779), of aristocratic Swiss stock, provided the detailed and accurate descriptions of glacial morphology in the course of alpine travels which helped found the sport of mountaineering, and provided Hutton with his imaginary geological excursions.

Chapter Three describes the views of Hutton and Playfair on alpine glaciation. They both agreed that the Jura erratics had been transported to their present position by glaciers over a formerly continuous slope from the Mt. Blanc Massif to the Jura, and both envisaged the application of glaciation to other mountainous regions, particularly as a mechanism for the dispersal of erratic boulders. In this they were decades ahead of general debate on stage four! Playfair's trip to the Alps and the Jura caused him to see that the formerly continuous surface was nothing more than the ice surface itself. Regrettably he died before the second edition of the 'Illustrations' was written. Without doubt a logical exposition of glaciation by Playfair might have an important effect on the historical development of geomorphology, given its intimate connections with the issue of drift and fluvialism. At about the same time as Playfair's tour, Perraudin, a Chamois guide, postulated an enlargement of the glaciers to account for the lines of stranded boulders high on mountain sides, above existing glaciers.

It is symptomatic of these early observations on glaciation that they remained stranded in their contemporary literature, significant but neglected, like 'les blocs erratiques' that they sought to explain. An important ancillary point is that up until 1862 glaciers were regarded merely as agents of transport giving rise to those deposits characterized as 'drift.' Their erosional ability was originally regarded as negligible. This view did not the neglect the erosional side of the 'erosion = deposition' equation, it merely assumed that all glacial debris fell onto the glacier from the surrounding valley sides; a view that a knowledge of alpine processes did nothing to discourage. Complementary to this attitude was the initial assumption by geologists that all glaciation was based simply on an extension of existing mountain glaciers, or upon existing mountain regions: an example of environmental bias that was not corrected until the limits of the North American ice sheet were traced, and the interiors of the Greenland and Antarctic ice sheets had been described.

The Jura erratics - 'quite as large as cottages'

A recurrent topic in the development of glacial ideas was the problem posed by the Jura erratics (Figure 9). The Jura mountains lie north and east of Geneva and are composed entirely of calcareous rocks, but their surface is littered with huge granite boulders, some even on the northern slopes lying in positions hidden from a direct line of sight to the Alps. Saussure, with his detailed knowledge of alpine geology, was able to trace these boulders to their source in the Mt Blanc massif some fifty miles away to the south east. Between the source and site of deposition lay a complex of valleys and smaller hills, to say nothing of Lake Geneva.

Saussure (1779) attributed them to the results of a great debacle, or flood, while de Luc (1790) thought they had arrived as a result of violent explosions. Hutton rejected both ideas, the latter on the basis that erratics also occurred in regions known to be completely devoid of any volcanic action.

Figure 9. *The Pierre-à-Bot, a granite block from the Mont Blanc range stranded above Neuchâtel in the Jura (from Geikie 1882a).*

Saussure (1796) also objected to explosions because the rocks should be shattered on landing, and ought to have produced craters, whereas his meticulous descriptions noted that the erratics rested on pedestals of limestone protected from solution by rainwater.

The erratics are of an immense size: one was estimated by Playfair to be 2520 tons, and the postulated agent of transport must needs match in power the problem in hand. They were not unique, erratic fields from the south Baltic coast were described by de Luc, and Hutton had noted them 'around all our granite mountains, and I believe all others' but their enigmatic relationship to the Alps invited speculative attention, and they were close to the tour routes frequented by the European intelligentzia. Saussure's debacle was revived in a more exact form by Sir James Hall (1815) in the paper reviewed in the last chapter. Hall used an ingenious geometric construction of the topography of the inclined plane which Hutton and Playfair had supposed to exist between Mt Blanc and the Jura. From this he could show that the granite outcrop would be permanently buried beneath glacial ice which he reconstructed by extending the present snowline out to its intersection with the inclined plane. He also pointed out that this extended ice cap still did not reach the Jura. Playfair (1802, #351), however, had invoked 'glaciers in the first place, and the torrents thereafter' as agents for the emplacement of the Jura erratics. Hall's reaction to torrents was:

> that the task is beyond the power of any river that flows on the surface of the earth; nay, it seems more than water, under any predicament, could accomplish, and more than could be expected from the debacle itself, however extravagant its magnitude may appear (p142).

I have already noted that Hall's judgement may have been defective here, but perhaps that was just as well for the glacial theory vis-a-vis floods, because Hall chose to use Wrede's suggestion that the erratics had been moved by 'floats of ice.' This was at least a step in the right direction and Hall connected it with the uplift of the Alps and the dislodgement of rocks and ice from their summits in consequence. It was far from a glacial theory, and it was still strongly catastrophic, but it was ancestral to Lyell's eventual scheme of marine submergence with iceberg rafted erratics. The Lyellian scheme persisted as an explanation of the Jura erratics, even with Ramsay, the principal glacial proponent in Britain, until the late 1850s.

A former extension of the glaciers a few miles beyond present limits was advanced by Venetz (1821), a Swiss engineer, but the suggestion fell on deaf ears. A later paper (Venetz 1829) made the much bolder suggestion that the Jura erratics were due purely to a former extension of alpine land ice, and that erratics in northern European localities also bespoke former glaciers, as independently suggested by Bernhardi (1832) who extended Polar ice southwards. Venetz's idea remained ignored until incorporated into Charpentier's better publicized views (1835) about more extensive alpine glaciers. In Norway, the only other European locality with existing glaciers, Esmark (1826) proposed a former extension of the glaciers to explain depositional features such as moraines, and the glacial erosion of over-steepened fjord walls. His 1826 paper was reprinted by Jameson in the 'Edinburgh New Philosophical Journal,' (1827) and, on evidence reported by Forbes (1846), was used in a lecture course by Jameson with references to the applicability of the glacial idea to Scotland. However, Jameson's notes to accompany the 1827 English edition of Cuvier's 'Theory of the Earth' still invoked debacles for the Jura erratics, and submergences elsewhere in northern Europe. Cuvier's views were, of course, totally catastrophic and diluvial, so that Jameson may have felt the need to adopt a sympathetic line when promoting his views in Britain.

The Jura erratics are diagnostic of attitudes towards glaciation because most authors discussed them with reference to their own theories. However, the glacial theory was soon to need a larger stage; one that would encompass the entire northern hemisphere, north of the Mediterranean. Henceforth the Jura erratics would be mere curiosities, relevant but incidental to a global refrigeration.

LOUIS AGASSIZ (1807-73): ADVOCATE OF AN ICE AGE

Agassiz made his international reputation as an expert on fossil fishes before he was thirty. He was converted to the glacial theory as a result of field excursions with Venetz and Charpentier in the summer of 1836. By the next summer he was ready to announce that a great hemisphere of glacial ice had covered the earth down to Mediterranean latitudes. The implication of this was far-reaching because it implied that areas now no longer possessed of active glaciers, Britain and eastern North America, should show the characteristic land-forms of glaciated terrain. Agassiz's initial papers connected the former ice with moraines and erratics, polished rock pavements and their associated striae. Both Agassiz and Charpentier obtained a British audience through the medium of Jameson's Edinburgh journal: no less than five papers in the period 1836 to 1839. It was probably upon these papers that Murchison (1839) based his rejection of the 'peculiar opinions' circulating on the continent about more extensive glacial ice, and its possible deposition of drift.

The first British convert to glaciation was Buckland,

once the biblical apologist, now in the vanguard of progress, who succumbed after initial stiff resistance during a tour of the Bernese Oberland in 1838. Buckland's undated private notes (published in North 1943) show that he was worried by the magnitude of the flood needed to move the Jura erratics. Von Buch had estimated they would require 'an impetus of 357 feet per second' and postulated the additional help of 'shocks.' Buckland also referred the inclined plane theory to Hall, rather than Hutton and Playfair. An earlier attribution by De la Beche (1832) of the Jura erratics to ice had had no effect, and may have derived from a careful reading of the 1822 reprint of Playfair (1802) together with Playfair's later notes, recorded in Playfair (1822). De la Beche was noted for his diplomatic handling of contradictory points of view. Buckland, in contrast, enthusiastically described to Agassiz the British glacial evidence that he recalled in retrospect, and invited him to Britain. After further Swiss studies in 1839, Agassiz came to Britain in 1840, presaged by a paper on polished and striated rock surfaces in the Alps (Agassiz 1840a). He recorded that:

> it was my wish to examine a country where glaciers are no longer met with, but in which they might formerly have existed (quoted in Davies 1969).

Agassiz gave four papers to the British Association Meeting in Glasgow, before making a Highland tour with Buckland. The result of the tour was a triumphant vindication of Agassiz's theory. The formal announcement that glaciation had indeed affected the Highlands was made in the newspaper 'The Scotsmen' on October 7th 1840, no doubt to the great delight of its editor McClaren, an amateur geologist who made his own contribution to glacial studies when he calculated that the massive hemispherical ice sheets demanded by Agassiz must result in sea level lowering of at least seven hundred feet, a figure which is the right order of magnitude.

One of the most critical sites examined by Agassiz and Buckland was the famed Parallel Roads of Glen Roy (Figure 10). Agassiz immediately recognized them as the shorelines of a pro-glacial lake impounded by glaciers in the Great Glen, exactly analogous to the Marjelensee on the Aletsch Glacier. The Roads had long excited speculation. Playfair (1816) mistook them for ancient irrigation works and, only the year before, Charles Darwin (1839) had published a long study concluding that they were marine beaches. Happily, this error did not deter Darwin from becoming a glacial convert, and two years later he was identifying glacial landforms in Snowdonia. However, his reclusive nature, and retirement to Down House in Kent, kept him away from the meeting rooms of the Geological Society where he might have influenced opinion more positively towards the glacial theory.

Agassiz toured Ireland alone and then returned to Scotland where he was taken to see Corstorphine Hill, Sir James Hall's field area, and which he agreed was the work of ice. Lyell, meanwhile, was converted to the glacial theory during an excursion with Buckland, near the family home in Forfarshire.

Figure 10. *The Parallel Roads of Glen Roy, viewed from the Gap (from Geikie's Scenery of Scotland, 1865).*

Agassiz returned to London elated that British evidence supported so well his scheme of huge ice sheets. A meeting of the Geological Society in November 1840 saw Agassiz, Buckland and Lyell ranged against the hostile scepticism of their assembled colleagues.

The glacial theory appraised

There were no immediate converts to the glacial theory in its pure sense. Murchison, who had been in the field with Agassiz in Scotland, remained unconvinced. Greenough was astounded at Agassiz's cool agreement that there had been at least three thousand feet of ice over Lake Geneva. However, Whewell drew attention to the diverse and widespread nature of the drift. The whole problem was confused by a number of issues. As Agassiz later admitted, 'it was in Scotland that I first acquired precision in my ideas regarding ancient glaciers,' and his papers and presentations tended to emphasize the sweeping generalities of the huge continental ice sheets he demanded. On the other hand, Buckland and Lyell described in minute detail particular field localities. In total, the two approaches added up to a very confused picture. There was no doubt that alpine glaciers had extended further in the past, and most people accepted, within a few years, that there had been analogous glaciers in various British mountains. The problem lay with the drifts. These deposits mantled lowland regions as well as highlands, and marine shells of arctic species had been found in Welsh drift to 1500 feet and above. Agassiz himself, in accounting for the end of the Ice Age, thought that the melting glaciers created a general marine submergence which resulted in bedded drifts and ice-rafted erratics. In this way he wedded Lyellian submergence with continental ice, and explained in a general way the existence of genuine till, bedded fluvio-glacial and outwash deposits, and the ever-present erratics. The confusing evidence of the shelly drifts in Britain, combined with the undoubted evidence of raised beaches with arctic fauna, provided every excuse for adopting a hybrid glacial theory which mixed large ice caps and general submergences. After Agassiz's departure, Lyell began to backtrack; the weight of the drift evidence, reinforced by North American evidence gathered in 1841/42, seem to have caused him to limit former glaciation to within a few miles of existing limits in the Alps, and to very restricted areas in British mountains. Lyell influenced world geological opinion through his books and lectures, and his recalcitrant position certainly slowed universal acceptance of the glacial theory.

One might characterize the glacial theory as constituting a 'slow catastrophe' and it is worth recalling that Agassiz remained to the end of his life what we would now call a 'creationist' in terms of his interpretation of fossil evidence in the geological column. Later in life he mistook the spheroidal weathering and the clays of tropical deep weathering in Brazil for glaciated terrain, a mistake that illustrates his tendency to generalize rashly. As Davies (1969) points out, his original concept of hemispheres of ice, based merely on a limited field knowledge in the Alps, was 'throwing scientific caution to the winds.'

Buckland's conversion to the glacial cause is also understandable in the sense that he too had always had catastrophic inclinations. Lyell was, in contrast, the uniformitarian and he may have sensed instinctively the conceptual dangers of the scheme, and the difficulties involved both in generating and dissipating the ice sheet within such a short period of recent geological time. Nevertheless, he himself was never specific about the source or timing of the general submergence he claimed, and which was no less a problem in energetics. Lyell eventually came to accept the bulk of the glacial theory, but it was a piecemeal conversion based on accumulating local evidence rather than any great change in philosophy. In a letter written in 1857 he admits that the complete lack of marine remains in the Swiss gravels, and the tracing of particular erratic trains (already summarized by McClaren 1842), forced him to accept a land-ice origin for the Jura erratics. Nevertheless, late editions of the 'Principles' still devoted more space to ice-rafted debris than to glaciation per se.

The glacial theory 1840-1862

In view of the lack of knowledge about the interiors of continental glaciers (Greenland and Antarctica) in the 1840s, and a corresponding over-emphasis on the arctic waters of northern Canada, it is understandable that most geologists could not match Agassiz's imaginative conception of continental ice sheets. In the period 1840 to 1862 most British and North American geologists remained wedded to a hybrid glacial theory involving highland glaciers and lowland submergences in varying proportions. In Switzerland and Italy, which lacked the tills admixed with dredged sea bottom shells that so confused discussion in Britain, the land-ice theory was accepted, and the former glacial advances were mapped in meticuluous detail. British geologists, long used to European field work, joined the fray. Forbes and Tyndall began a long series of investigations into glacial motion which amplified the initial studies made by Agassiz (1840b, 1847), and although they had no immediate impact on the acceptance of the overall theory, they had a cumulative effect by providing a basis of understanding for glacial mechanics. The influence of Von Buch and Humboldt, still conservatively catastrophic, was sufficient in Germany to forestall any acceptance of the land-ice theory, and the glacial origin of the north German erratics was, amazingly, not accepted in Berlin until about 1875.

Lyell left for his first North American trip, lasting thirteen months, in the summer of 1841. He noted very carefully the southern limits of the drift, observed roches moutonees in Canada, recorded a disappointing lack of ice polishing on waterfall crests in Canadian rivers and was particularly anxious to use the new terrain as a test of the glacial theory, construed broadly as very localized mountain glaciation and extensive marine submergence. He found a lack of sufficiently large mountains for glaciers, but bountiful evidence of ice-rafted boulders, even in the sea-ice around Nova Scotia. His influence caused Hall (1843) to assess his New York evidence in the light of the submergence theory, although its deficiences became more apparent to him as the field evidence accumulated.

Agassiz arrived in North America to lecture in 1846, and to settle shortly afterwards. His theory preceded him because Hitchcock (1841) gave an able summary in which he drew attention to the similarity of the phenomenon in question on both sides of the Atlantic, and the likelihood that they had a common explanation. However, he still thought that an extensive ice cover required mountains and was worried about apparent glacial deposits far from any obvious source. McClaren's lucid review (1842) of Agassiz's theory was reprinted by 'Silliman's Journal' for the North American audience, and included his remarkable estimate of sea level lowering (p124), and a sceptical view of the proposed uplift of the Alps with which Agassiz terminated the Ice Age.

The shift towards an acceptance of the land-ice version of the Ice Age was caused by the cumulative effects of several factors. Perhaps primary amongst these factors was the field evidence itself. Geology, in general, was moving apace and even if British geologists were resistant to having a foreign theory pushed upon them they had long been obedient to demands for a rational interpretation of field evidence. The detailed mapping of local areas, and the increasing attention being paid to stratigraphic and sedimentological considerations, forced geologists to acknowledge that extensive land-ice provided more answers than submergences, with their submarine currents possessed of arbitrary powers. For example Forbes (1846) wrote of the Isle of Skye:

> Having hitherto taken no share in the discussions raised as to the proofs of glacial action in this country, and being as indisposed as ever to embark in a theory which offers such evident difficulties, I yet feel it to be a duty to make known what I have observed in connection with it among the Cuchullin Hills; phenomena so singular and well marked as to require a steady and patient consideration in whatever way they may be attempted to be explained, and which I am compelled to admit, (whatever geological causes, now unsuspected, may hereafter be discovered,) must now be unhestitatingly ascribed to the action of moving ice, rather than to any other kind of agency with which we are acquainted.

He then described in graphic detail glacially moulded rocks, striae, and perched blocks around Loch Coruisk, and remarked that he had seem similar features elsewhere in the Cuillins, and in the Alps. Lyell yielded, reluctantly, over the Jura erratics, and Murchison came to accept by 1862 that his native British mountains had been glaciated, although not actually eroded.

Another facet of this increasing field experience was the slow recognition that alpine glaciers might have accomplished very significant vertical erosion. Ramsay was originally anti-glacial, but by 1851 Mantell could record in his diary that he had heard at the Geological Society 'a wild dream about glaciers and the glacial theory by Prof. Ramsay.' Ramsay generally adhered to the hybrid theory, but in 1859 he 'explained the glacial origin of certain rock-basins now holding lakes, on the watersheds and in the old glacier-valleys' of both North Wales and Switzerland. At the same time he still believed that a marine submergence had been responsible for rafting the Jura erratics across the Swiss Valley from alpine glaciers terminating in the ocean. Further field excursions eventually persuaded him that this was untenable: no marine drift could be found between the Alps and the Jura, and no drift at all could be found on the wide table-lands to the north. He then showed that the Swiss and Italian valleys owed nothing to fractures or gaping fissures and that they were over-deepened exactly where the ancient ice would have been thickest. By reconstructing the Rhone glacier and Lake Geneva in a scale cross-section he revealed that the excavation of the lake basin was not as dramatic as it appeared to be. On the basis of these findings he had no hesitation in extending this explanation to all northern hemisphere lake basins where they were surrounded by other evidence of ancient glaciers (Ramsay 1862):

> I am therefore constrained to return, at least in part, to the theory many years ago strongly advocated by Agassiz, that, in the period of extremest cold of the Glacial Epoch, great part of North America, the north of the Continent of Europe, great part of Britain, Ireland, and the Western Isles, were covered by sheets of true glacier-ice in motion, which moulded the whole surface of the country, and in favourable places scooped out depressions that subsequently became lakes.

He was able to report in a footnote that Sir William Logan, Director of the Geological Survey of Canada, had independently reached the conclusion that the Great Lake basins were of glacial origin. He felt able to extend his interpretation of the rock basins of North Wales to include the larger valley bound lakes such as the 'Lakes of Llanberis and Llyn Ogwen.' The Lake Geneva problem was now solved, the limnological objection was removed from the fluvial theory and the coast was clear for Jukes's paper, published in the same year, to provide a rational basis for subaerial landscape evolution. Jukes himself observed that 'the formation of lakes lying in "rock-basins" .. had always been a complete puzzle until I read Professor Ramsay's paper' (quoted in Davies 1969).

Ramsay's paper was of great significance in that it attributed to glaciers the power to create erosional landscapes, in contrast to the previous belief that they merely occupied pre-existing valleys. However, Ramsay did not banish submergence completely. In a rude chronology of the British glaciers, developed at the end of the paper, he envisaged a period of 'great original glaciers' which probably gave rise to the 'unstratified boulder-clay,' a period of submergence when the glaciers had retreated somewhat and the 'stratified erratic drift' was deposited, followed by a period when the glaciers 'were enlarged again during the emergence of North Wales and other countries so as to plough the drift out of many valleys .. the newer development of glaciers was strictly local.' Realizing the environmental implications of glaciation he found it:

> difficult to believe that the change of climate that put an end to this could be brought about by mere changes in physical geography. The change is is too large, and too universal, having extended alike over the lowlands of the Northern and the Southern Hemispheres.'

Thus the submergence was further circumscribed by evidence
and reduced to an increasingly specific role. The spectacular
raised beaches around western Scotland and the shelly drifts
of Wales still awaited explanation. The latter helped
perpetuate the notion of partial submergence, deeper than the
beaches indicated, even into the twentieth century. It is
interesting perhaps that Ramsay refers to 'the time of
depression' which reveals that he was thinking of a relative,
rather than an absolute, submergence; possibly foreshadowing
the notion of glacial isostasy, but probably devised ad hoc
to produce a local British submergence, independent of the
non-marine Swiss drifts.

AGASSIZ VINDICATED: ICE ALMOST EVERYWHERE

The glacial theory received valuable publicity through the
interpretation of British landscapes, with a particular
emphasis on glaciation, by two influential officers from the
Geological Survey. Ramsay's popular book on 'The Physical
Geology and Geography of Great Britain' first published in
1863 ran through six editions and included both his glacial
theory and his adoption of fluvialism a la Jukes, with
submergences restricted to a minor, and mainly depositional,
role. A complementary volume on 'The Scenery of Scotland'
(1865) by a young colleague of Ramsay's, Archibald Geikie,
had a similar philosophy. Both authors had rediscovered
Hutton and Playfair and were amazed at their insight, so long
submerged in speculative floods.
    Geikie (1863) made his own contribution to the glacial
theory through a long and detailed paper on the structure of
Scottish drifts which he showed conclusively could not be
submarine deposits. However, it was James Geikie, a younger
brother of Archibald and also an officer in Survey, who wrote
the first definitive text on the Ice Age.

James Geikie (1839-1915) and 'The Great Ice Age'

James Geikie joined the Scottish Geological Survey in 1861
and used the opportunity provided by field mapping to collect
information on all aspects of glacial landscapes, together
with what he could glean from quizzing colleagues working in
other areas. Geikie used the material, together with a
comprehensive coverage of foreign literature, to write 'The
Great Ice Age' (1874) which was subtitled 'and its relation
to the antiquity of man' and dedicated to Andrew Ramsay.
    The book, which was published simultaneously in the
United States, gives 'a systematic account of the Glacial
Epoch, with special reference to its changes of climate' as
the preface notes. There is a strong emphasis on the Scottish
and English evidence, but there is material on Ireland,

Scandinavia, Switzerland, Northern Italy and North America. The first edition retained the notion of a general, deep, post-glacial submergence, but by the second edition (1877) Geikie had adopted Jamieson's view that post-glacial submergence was limited to 100 feet above present sea level, as evidenced by the raised beaches, although he was prepared to admit over 500 feet during an inter-glacial on the evidence of marine clays at this altitude. On the related topic of rock-basins a revealing footnote in the second edition shows the extent to which influential resistance was still being encountered:

> Sir C. Lyell's hypothesis is so manifestly inapplicable to the rock-basins of Scotland that in the previous edition of this work I passed it in silence, and this I was more inclined to do as Sir Charles himself never tried to apply it to the Scottish lakes. I hardly thought that any geologist could indulge the hope that "unequal movements of upheaval and subsidence" would account for the phenomena in question, but I have been mistaken; hence the above paragraphs (Geikie p277).

One aspect of the first edition that developed in later editions was the suggestion that drift stratigraphy implied more than one period of ice advance, with an inter-glacial period between. The implication of the first edition, made partly to resolve conflicts between the post-glacial fauna of Scotland and the faunal assemblages of the Palaeolithic gravels of England, became assertive in the second edition:

> In every country where glacial deposits have been studied with any attention to details, we have clear and convincing proof of a mild interglacial period having supervened in what one may term the later stage of the glacial epoch (Geikie 1877, p491).

One may note the mental reservation on the length of the glacial period, and he later speaks of the 'last interglacial period.' In southern England he equated the Palaeolithic gravels, in which there were traces of pre-historic man, with the inter-glacial. The topic of early man was important because of the controversy still surrounding Darwin, Wallace and the theory of evolution.

An underlying influence in all editions of Geikie's book was his friendship with James Croll (1864, 1875) who had devised an astronomical theory from which it might follow that 'although a glacial climate could not result directly from an increase in the eccentricity of the earth's orbit, it might nevertheless do so indirectly.' Geikie, in seeking a controlling theory for the Ice Age, thought that Croll's work provided the most satisfactory rationale, despite its mixed reception from the astronomers.

The book was well received, at home and abroad, and was probably particularly influential in Germany (where according to Newbigin and Flett (1917), glacial studies were still relatively backward), Switzerland, Sweden and the United States. The emphasis on the total Quaternary environment, the careful and detailed analysis of drift stratigraphy, and the strong assertion of all the essential elements of Ramsay's erosional views came at just the right time to stimulate investigation, suggest hypotheses, and provide a valuable worldwide perspective on glaciation. Its eventual importance can be judged from the facts that Mt. Geikie was named for its author, in Wyoming in 1900, and the monumental work of Penck and Bruckner (1909), which established multiple glaciations in the Alps, was dedicated to him.

The third and final edition was published in 1894 after very substantial revision and the incorporation of chapters on North America especially written by Chamberlin, of the United States Geological Survey. After assessing all the evidence Geikie went much further than in the second edition and proposed the view that there had been six glacial periods, with corresponding inter-glacials. Newbigin and Flett (1917) record in their biography of Geikie that:

In foreign countries generally he found more support, though everywhere, it may be admitted his views must have been regarded as extreme.

Most British geologists had scarcely dried their feet after an epoch of submergences and were resistant to the idea of multiple glaciations, for which, to be fair, the native evidence was poor. However, in the Alps and in North America more complete sections favoured the hypothesis of multiple stages. From the point of view of landforms the multiple-stage hypothesis made it possible to see bedrock excavation by glaciers in a more reasonable light: repeated phases and inter-glacial weathering lighten the erosional load on any single advance and made a slow catastrophe even slower. It also made possible a differentiation of landforms and drifts on the basis of age, where variations in morphological 'freshness' on otherwise similar materials had previously been confusing.

Geikie's book, in its three editions, established the Ice Age in the international literature for at least three decades after 1874, with a powerfully argued orthodoxy, prophetic of, and central to, subsequent work. Adherents to substantial glacial submergence still remained. Dawson (1893) in Canada promoted it until his death in 1899, and in Bonney and Garwood a small school of glacial 'protectionists,' inspired by the Alpine geologists Rutimeyer and Heim, argued that glaciers protected landscapes from bedrock erosion.

There remained some unanswered questions, some puzzling relationships, whose significance escaped most, but not all, nineteenth-century glacialists. These problems centered on the problem of relative and absolute changes of sea level.

Unresolved questions: glacial eustasy and isostasy

McClaren (1842) pointed out that Agassiz's large ice caps would necessarily imply the abstraction of water from the oceans. He computed the drop at about 700 feet and thought that remaining land ice would cause an additional rise of 100 feet. The estimate is astonishingly good considering the imprecision of Agassiz's original scheme.

However, in an era when submergences contemporary with real or supposed glacial ice ruled supreme, there was little theoretical call for a notion that would drop sea level rather than raise it. McClaren himself did not foster the idea since in 1866 he was still supporting a version of the hybrid theory of glaciation, and the shell species in the raised beaches were arctic, indicating that they were roughly contemporaneous with the end of the glacial period. The idea re-appeared with James Croll (1867), an early believer in the land-ice theory, who regarded it as one component in his theory that the mass of the ice sheets affected the earth's centre of gravity and so shifted sea level. Tylor (1868), another independent thinker, proposed a 600 drop in sea level during the glacial period and looked for evidence in the deep bedrock gorges possessed by most rivers, in the basal deposits of deltas such as the Mississippi, and in the development of coral islands. Once more the idea fell on barren ground for in the 1860s Ramsay's paper on glacial erosion was only adopted by the avant-garde and submergence, to some degree, was still popular. The problem of the raised beaches remained.

The complementary idea that would eventually help to resolve the problem was proposed by Jamieson (1865, 1882) and Shaler (1874) who were convinced that the great weight of the ice sheets would depress the earth's crust. This idea depended on accepting a model of earth structure that involved a thin, but rigid, crust overlying a 'liquid' interior. A rival theory held that the great cold at the base of the ice cooled the crust and caused crustal shrinking and effected depression by this mechanism. In either case, general acceptance was likely to hinge on the opinion of physicists. The topic was an actively developing research field in geophysics and the relevance of the geodesic work on the Himalayas and the emerging theory of isostasy was only slowly becoming apparent. In addition, geologists were actively battling geophysicists over a much more important topic: the age of the earth and its impact on the rate of geological processes. The isostatic idea arose separately in the United States, where Dutton (1889) proposed it in the context of erosional unloading, and Gilbert (1890) considered the isostatic deformation caused by ancient Lake Bonneville.

It occurred to nobody that the two ideas, working together, could usefully explain late-glacial, arctic water, raised beaches in northern Europe. Geikie (1892) summarized the subject for the British Association:

> The strand-lines in high latitudes, however, are certainly connected with glaciation in some way not yet understood; and if it cannot be confidently affirmed that they indicate regional movements of the land, the evidence, nevertheless, seems to point in that direction.

Similarly, in a review of Nansen's scientific work in Greenland, he asks whether ice sheets can push down land surfaces, and have them rise again on melting? Can they gravitationally attract the ocean by virtue of their mass, and can ice sheets shift the earth's centre of gravity? Each question he says:

> has been answered in the affirmative and the negative by controversialists, and, until the geological evidence has been completely sifted, each, doubtless, will continue to be affirmed and denied. All that need be pointed out here is that some of the movements which occurred during the Pleistocene period were on much too large a scale to be explicable by any of the hypotheses referred to (Geikie 1891).

It is fitting to take leave of the Ice Age on this note of query, and the appeal for more field work. Daly, a geophysicist, beginning in 1910, brilliantly revived and integrated the eustatic and isostatic theories by taking a global view of the evidence and using it to test models of earth structure which reproduce it. Penck and Bruckner (1909) confirmed Geikie's daring extrapolations by finding four alpine glaciations. The minutia of glaciation were by no means settled in the early 1900s, but the basis of our present understanding was established and to promote the understanding of modern glaciers, used as a basis for interpreting those of the Pleistocene, Hobbs (1911) wrote his 'Characteristics of Existing Glaciers.' Thus, while Geikie was still alive, the Ice Age came of Age.

Ideal section across inland-ice.

Chapter Eight

NORTH AMERICAN GEOMORPHOLOGY UP TO 1900

I have wished for a sight of that gap, through which the rivers,
gathered in the long valleys of those mountains, break through
the ridge and find a passage to the sea.

(James Hutton, 1795)

Had the birthplace of geology lain on the west of the Rocky
Mountains....the efficacy of denudation....would have been one
one of the first obvious principles of the science.

(Archibald Geikie, 1882b)

Long after the American War of Independence, North America
was a net importer in the realms of culture and science. It
did, however, export travellers' accounts of its landscapes
and these in turn served to stimulate geological thought at
home and abroad. Niagara Falls, the Great Lakes, the
Mississippi, the ordered ridges of the Appalachians, the
endless frozen North, both land and ocean, and eventually the
varied terrain of the Rockies; all these, at various times,
gave rise to geological speculation and new hypotheses.
    There were no schools of indigenous geologists in North
America prior to the nineteenth century. White (1978) termed
Lewis Evans, the best of the eighteenth-century analysts, 'a
proto-geologist.' Evans was best known as a cartographer and
surveyor but his 'Map of the Middle British Colonies' (Evans
1755), accompanied by a thirty-six page booklet of 'Analysis
..,' contained a wealth of geological information that was an
unacknowledged source for later writers for up to sixty
years. White (1951a) wondered if Hutton ever saw the Evans
map. He may have, because, as I noted in Chapter Three in
connection with the Appalachian water gaps, Hutton made it
clear that he had studied American maps in detail and Evans'
maps of Pennsylvania (1749, 1752) contain the prophetic
annotation that the Appalachians 'furnish endless Funds for
Systems and Theories of the World; but most obvious to me
was, That this Earth was made of the Ruins of another.' Even
more extraordinary was his explicit suggestion (Pownall 1776,
Klinefelter 1971) that:

    the highest Mountains themselves, as they now appear,
    were formerly but one large Plain, inclining with a
    considerable slant to the sea; that this has been worn
    away into its present Appearance of Ridges, with vales
    between them, by the Rains of the Heavens and the Waters
    of the Earth working away the Soil from the upper parts,
    and carrying it down Seawards. That the Soil thus
    carried down and lodged in various places hath in a

Series of Ages formed the lower Plains of the Jerseys, Pennsylvania, Virginia and the Carolinas.

He proposed this, apparently from his own thinking since no other source is known, together with an alternative, that the Appalachian topography was cut by a flood, of which the Great Lakes were a remnant. Evans apparently preferred this version although he did think that the power that Dr Woodward 'ascribes to the water of the Deluge .. too much of a Miracle to obtain Belief.' White (1951b) also found in the writings of Evans an early reference to the concept of isostasy, proffered as an explanation for the uplift of mountains containing fossils:

> This part of America, disburthened of such a Load of Waters, would of course rise, as the immediate cause of the shifting of the Center of gravity in the Globe at once or by Degrees, much or little, according as the Operation of such Event had Effect on that Center.

However, the idea is less like isostasy than Hooke's notion (1705) of shifts in the planetary axial tilt.

I reviewed in Chapter Five the development of ideas on Niagara Falls, and the way that they reflected contemporary patterns of thought. The 'Heroic Age of Geology' at the beginning of the nineteenth century in Britain and Europe saw only pale reflections in North America, and in the study of landforms there was no recognition of the work of Hutton and Playfair; diluvialism ruled supreme, although it had to work hard to overcome the prejudice of local opinion about the origin and development of Niagara Falls! The rejection of fluvial notions was not made entirely in ignorance. Mitchill (1764-1831), an early New York geologist, was a foreign member of the Royal Society of Edinburgh from 1793 and was therefore likely to be informed about the Huttonian school. He also contributed an appendix on the geology of North America to Jameson's translation (1815) of Cuvier's 'Essay on the Theory of the Earth.' Benjamin Silliman, who founded in 1818 what is now 'The American Journal of Science,' a primary source for geological writings on North America, went to Edinburgh from 1805 to 1806 and heard Playfair, Jameson, Murray and Hope lecture on all sides of the great debate. In 1829 he adopted the third English edition of Bakewell's 'Introduction to Geology' as his course textbook in Yale, and added his own extended appendix with heavily diluvial and catastrophic overtones. Most British and continental texts became available in North America, but none before Lyell gave much credence to fluvial ideas.

In summary, therefore, the Atlantic filtered heretical ideas out of British geological writings, allowing only the more conservative views to survive the transplantation.

THE AGE OF LYELL: 1830 ONWARDS

The appearance of Lyell's 'Principles of Geology' after 1830 coincided with the establishment of the geological surveys of the eastern States during the 1830s. Thus the book destined to be the most influential geology text of the century was available just as the needs of ground survey forced groups of professionals into the meticulous attention to detail that should be the precursor to all theorizing.

Lyell himself came to visit the United States and Canada for thirteen months in 1841/2. The financial underpinning was provided by a series of public lectures on geology that were given in Boston, New York and Philadelphia, to audiences of several thousand. The rest of the time Lyell devoted to 'geologizing' in extended tours during which he met almost all the significant figures in North American geology including James Hall of the New York State Survey, Benjamin Silliman, Professor Eaton of Troy, Professor Rogers of the Pennsylvanian Survey, and Professor Hitchcock of Amherst College, Massachusetts. His influence on James Hall I mentioned in Chapter Six, and his advocacy of the iceberg theory, with very limited mountain glaciers to provide the icebergs, was a powerful factor in delaying general acceptance of the full glacial theory for many years.

His visit had the effect of legitimizing the labours of American geologists for the European audience. In turn it provided an authentic presentation of the doctrine of uniformity at a time when the results of the State Surveys east of the Mississippi cried out for a theoretical fabric. The battle for uniformity, as an overriding principle, was largely won in Britain by 1840, and there was nothing in the State Surveys that suggested that it need be fought again in North America. However, as in Britain, skirmishing still remained on the issues of fluvialism and glaciation.

Lyell published thirteen papers on the basis of his American trip, and his two volume 'Travels in North America' was published in 1846, just as he returned for another trip. This time the focus was on the southern United States, and in particular on the Mississippi river, a prime exhibit for the fluvialist case. Lyell was more interested in the character of the delta and its accumulation over time, than the fluvialist case as such, but the evidence for uniformity was telling, whatever the precise focus of interest, and he calculated an age for the delta upwards of 67,000 years (Lyell 1847). As with his first visit, Lyell's observations became enshrined in subsequent editions of the 'Principles' where they provided invaluable food for thought for students at all levels. In this way, American evidence became an essential part of European geological thinking and paved the way for the eventual acceptance of modern fluvialism.

The Mississippi, by virtue of its very size, provided

excellent data for assessing the rate at which land surfaces are degraded. As the nineteenth century progressed, economic exploitation of the Mississippi basin for wood, coal, oil, and agriculture ensured that the river carried an increasing burden of traffic. In consequence, formal studies of its nature and behaviour were authorized by Congress in 1850 which culminated in the influential report by Humphreys and Abbott (1861). Here, in comprehensive detail, was the hydrology of one of the largest river basins in the world, and it provided evidence at every level from the details of river flow, to the rate of growth of the delta. Their refined analysis included estimates of traction load and suggested a figure twice that estimated by Lyell for the annual sediment discharge. Archibald Geikie seized on the report as an invaluable tool in the fight for fluvialism and used it in an important paper on rates of erosion (Geikie 1868) in which he estimated that the continents would be worn down to sea level in between four and seven million years, and that marine erosion was orders of magnitude slower.

NORTH AMERICAN INDEPENDENCE

State Geological Surveys and Federal initiatives

The development of a genuinely independent American geology began with the experience gained in the various State Surveys of the 1830s and their sporadic successors in subsequent decades. In New York James Hall (1843) wrote a report including material on landforms and drift, together with a discerning assessment of contemporary theories that was as good as anything in Britain at that time. The broad perspective that came from covering the entire state was possibly of more immediate value than the detail forced on the British geologists who were constrained to work within the arbitrary limits of Ordnance Survey one inch sheets. However, when the Survey was finished Hall continued only with palaenontology, and the impetus was lost. Edward Hitchcock was another who worked on a State Survey, that of Massachusetts in the 1830s, and it was he who gave the glacial theory its initial public airing in North America.

The only Federal enterprises were occasional expeditions to areas west of the Mississippi, such as Featherstonehaugh's reconnaissances (1834 in the Ozarks, 1835 westwards to Wisconsin), and a decade later the four year Wilkes Exploring Expedition around the Pacific to which James Dana was attached. Dana (1849a,b) made good use of his trip in the same tradition as Darwin and published in 'Silliman's Journal' descriptions of Hawaii and Australia that called only upon subaerial erosion to produce the deeply dissected terrain (Dana 1849b). Dana had no doubts about the efficiency

of surficial processes and called upon the point proved by
Hopkins (1844), that 'the transporting power of flowing water
increases as the sixth power of the velocity,' in association
with large and frequent 'rises' of the streams. He sketched
the evolution of the long profile of a river at various
intervals of time and noted that an eventual balance between
erosion and deposition would 'determine a permanent height
for the bottom of the lower valley.' In a review of Chambers'
'Ancient Sea Margins' (Dana 1849c), he pointed out that not
all terraces were necessarily marine and noted that
instrumental surveys were needed to establish exact
longitudinal slopes, and that a relative drop of sea level
would create river terraces, as well as raised beaches.

Dana made further use of his experiences, and he had
also cruised in the Mediterranean, when he wrote his 'Manual
of Geology' (1863), widely regarded as the best textbook on
geology since Lyell's 'Principles.' He was an early believer
in the land-ice version of glaciation, a complete fluvialist
who quoted Du Buat, and one whose views on the development of
landscapes were well in advance of their time.

At about the same time, a young civil engineer in Canada
was writing a report on the remedial steps needed to prevent
the silting of Toronto Harbour that could only win second
prize in the Harbour Commissioners' competition of 1854, but
which G.K.Gilbert (1890) would term a 'classic essay' only
thirty five years later. Sandford Fleming was a Scottish
expatriate who did his own surveying and engraving. His entry
was based on a paper already written in 1850 and which
contained an astonishingly complete introductory essay on
beach processes as a systematic basis for his practical study
of the harbour.

His grasp and application of uniformity was mature and
confident; one diagram predicted the future shape of the
Toronto Peninsula spit, yet, an incidental reference to 'the
very apparent marks of design' suggest more than a passing
acquaintance with his illustrious compatriot, James Hutton.
In a later paper (Fleming 1861) he identified the Davenport
gravel drift, then west of Toronto, as a spit formed in an
identical fashion to the then Toronto Peninsula, when the
level of Lake Ontario was higher, i.e. the pro-glacial Lake
Iroquois of modern literature. Sandford Fleming went on to
become Canada's greatest nineteenth-century civil engineer.

One other notable Canadian finding was the view of
William Logan (1863), Director of the Geological Survey of
Canada who confirmed the insight of Ramsay (1862):

> These great lake basins are depressions, not of
> geological structure, but of denudation, and the grooves
> of the surface rocks which descend under their waters
> appear to point to glacial action as one of the great
> causes which have produced these depressions.

These advanced views were actively combated for thirty more years by J.W.Dawson (1893) who continued to believe in glacial submergences in the style of Lyell. Curiously, Agassiz's presence in North America after 1847 seems not to have accelerated acceptance of land ice glaciation, and although he did field work in the Great Lakes, New England and Brazil his impact was pedagogical, not theoretical.

Federal Exploration in the American West

Exploration west of the Mississippi developed slowly and sporadically, and usually under the aegis of military expeditions looking for suitable railroad routes westwards. Before the Civil War the two names of importance are Newberry and Hayden. Newberry (1862) saw clearly in the Grand Canyon region that 'the broad system of valleys..belong to a vast system of erosion' which was undoubtedly subaerial. A later expedition undertaken in 1859 remained unpublished, due to the Civil War, until 1876 and, while its main conclusions merely re-iterated the earlier ones, there is one interesting statement that may have attracted the imagination of the youthful and wide-reading W.M.Davis:

> Even the imagination itself is lost when called upon to estimate the cycles on cycles during which the much grander features of the high table-lands were wrought from a plateau which once over-spread most of the Colorado Basin, burying the present Sage-plain 2000 feet beneath its upper surface (Newberry 1876).

Hayden (1862) was similarly impressed by the power of slow erosion which he linked to that of slow uplift of the mountains in a description of the condition that was later termed 'antecedence' by Powell (1875).

After the Civil War western exploration, with Federal funding, proceeded apace and the four main surveys were led by King, Hayden, Powell and Wheeler. Powell's descent of the Colorado through the Grand Canyon in two separate trips (1869 and 1871/2), and the subsequent published account of the journeys' fired the public and the scientific imagination. Here, in unparalleled splendour, far from the complications of surficial drifts and raised beaches, was unquestionable evidence of the power of ordinary rainwater, given time, to slice through thousands of feet of rock. There could be little doubt, according to explorers' accounts, that the river could remove the detritus supplied to it from above.

The scientific results of the Colorado expeditions and related trips are contained in Powell (1875, 1876). The modern significance of his work hinges on three points: the nature and potency of erosion, the idea of base level, and generic classifications of landforms. At the time, the

detailed discussion of weathering and erosional processes and
their relationship to the gradual evolution of valley systems
with their associated slopes was of undeniable importance. It
can, however, be overstressed since with the work of Dana in
the United States, Jukes, Ramsay and Geikie in Britain,
Rutimeyer in Germany, and the popularization of Reclus (1871)
in France, already fluvialism had reasserted itself as an
explanation of landscape. Nevertheless, the American evidence
was welcome, as the symbolic adoption of the Grand Canyon as
a frontispiece in British textbooks, made clear, (Figure 11).
Archibald Geikie (1882b) was so impressed by the reports that
he visited the American West in 1879 as a guest of the
newly-formed Geological Survey (U.S.G.S.).

THE PLATEAU AND CAÑONS OF THE COLORADO.

Figure 11. *Part of the frontispiece from Geikie's "Textbook of Geology" 1882
Drawing by W.H. Holmes, a colleague of Powell.*

Powell's two other findings were of greater long term import. He distinguished three types of river system: that which is 'consequent' on the nature of the initial geological surface, that which becomes 'superimposed' by erosion onto a different geological structure, and that which is 'antecedent' to a rising geological structure, but which manages to maintain its path against the developing uplift. These ideas were adopted by W.M.Davis as part of his Geographical Cycle together with another organizing principle that Powell developed: the concept of base level.

Base level, as defined by Powell (1875), was a limit to progressive denudation such that 'no valley can be eroded below the level of the principal stream, which carries away the products of its surface degradation.' Thus the sea, a main river, or any hard rock outcrop in a stream could act as a local and temporary base level. In general it was an ideal surface below which, in a given region, river erosion could not reduce a landscape. By comparing a stream to this ideal he could state that:

> the more elevated any district is, above its base level of denudation, the more rapidly it is degraded by rains and rivers (Powell 1875).

Other classifications developed by Powell have less modern interest. The very clarity with which geological structure was revealed in the semi-desert landscapes of the West seduced Powell into a classification of drainage systems based on the nature of the underlying structure: anticline, syncline, dome, basin etc. Many of the first textbooks on geomorphology adopted a strongly structural approach; James Geikie in his 'Earth Sculpture' (1898) and John Marr in his 'Scientific Study of Scenery' (1900). The appeal is natural to the trained geologist since it welds the scenery to the underlay, but it effectively negates the idea that a stream system is an integrated unit operating under general laws by focussing on the infinite variety of patterns that will result from its inevitable response to structure.

G.K.Gilbert: the geomorphologist's geomorphologist

Gilbert (1843-1918) began his geological career on the Newberry survey of Ohio (1869), transferred to Wheeler's western survey (1871), and in 1875 joined Powell. Shortly thereafter he began a study of the Henry Mountains which was completed in two months and published in 1877. The fifth chapter in this report, which has been widely cited ever since, restates 'certain principles of erosion which have been derived or enforced by the study of the Colorado Plateaus.' It is comprehensive, yet terse, a 'Euclidean essay in plain English' (Chorley and Beckinsale 1980). Gilbert used the skeletal landscape to expose the mechanics of weathering, transportation and corrasion, and their eventual effects on the landscape: graded slopes leading to lateral planation. Gilbert's approach was always that of the engineer: sound mechanical reasoning of a qualitative nature backed up by approximate calculations at crucial points. His underlying rationale was that the landscape achieved over time a state of dynamic equilibrium and that in a landscape 'there is an interdependence throughout the system.' His most influential statement, although anticipated by Dana (1849a,b), was that 'downward wear ceases when the load equals the capacity for transportation' so that 'lateral erosion becomes relatively and actually of importance' with the result that landscapes become planated across geological structure. These conclusions were crucial to Davis' development of the Geographical Cycle and in this way Gilbert contributed unwittingly to a model that he himself never used.

The attraction of Gilbert to the modern scholar lies in his clear and precise approach to complex problems and his use of the steady state model: that things have always been much as they are now. His emphasis on equilibrium led him to consider, for example, the forces which were responsible for upwarping the shorelines of ancient Lake Bonneville which he had carefully levelled (Gilbert 1890), and to conclude that they reflected the recovery of the crust from the load imposed by lake waters equivalent in weight to 730 cubic miles of rock. His pioneering study in isostasy was ignored for seventy years, although his colleague Dutton (1889) was already proposing the same principle for erosional unloading of the earth's crust, and Chamberlin, the glacial geologist, was also testing the idea that the Pleistocene ice sheets had depressed the earth's crust, an idea that Gilbert admitted using as 'a working hypothesis.'

When the United States Geological Survey was formed in 1879 to rationalize the many Western Surveys, Gilbert soon became an assistant to Powell and his field activities were relegated to spare time and leaves of absence. A projected study of the Great Lakes never materialized although he did manage to salvage some perceptive studies on Niagara Falls; both its history and its rate of retreat (1895, 1907). His work slowly revived after the death of Powell in 1902 and his final years were devoted to work on the transportation of debris by flowing water, studied in flumes, and its application to the problem of hydraulic mining in the Sierras behind San Francisco (1914, 1917). He was disappointed when his flume studies failed to reveal a unified set of controlling equations for he found that 'the actual nature of the relation is too involved for disentanglement by empiric methods.' This later work was far too advanced, technically, to be assimilated into a subject that was rapidly becoming more, rather than less, qualitative under the influence of

W.M.Davis and his disciples.

Gilbert made important contributions to many branches of geomorphology (Yochelson 1980) and was greatly respected by his contemporaries, but he failed to develop a Gilbertian school during his lifetime, in part because he never took up a University appointment. In recent years he has been resurrected in his own right, instead of being revered through darkened Davisian spectacles. His appeal to a subject now addicted to systems theory and a medley of equilibrium states is obvious, but a more enduring reason is that 'he preferred the study of modern processes to the study of ancient Earth history' (Baker and Pyne 1978): exactly the emphasis of modern work in the last two decades.

The integral view: W.M.Davis and the Geographical Cycle

Geomorphology at the end of the nineteenth century, like an uncompleted jigsaw, awaited a visionary thinker who could stand back and see the whole. William Morris Davis was that man, for better or for worse. Taking the pieces provided by the Western Surveys, especially the work of Gilbert on graded, planated slopes in the Henry Mountains, the complementary idea of base level from Powell, and the assumptions of the new diastrophism (see p117) Davis, stimulated by a summer in Montana (1883), and the threat of losing his Harvard job, produced the Geographical Cycle. In essence it said that subaerial erosion would wear down a landmass through a predictable series of stages to a surface of ` low relief, unless 'interruptions' or 'accidents' intervened. In essence, therefore, it said nothing that enlightened thinkers had not deduced already, several times, during the last two and a half millenia.

There was, however, a very significant difference. The nature of the accidents were specified precisely now; glaciation, climatic change, vertical movements of the crust (epeirogeny), and volcanism. A working terminology was provided, a sequence of morphological stages was deduced, and the whole lay within an acceptable general framework of geology. The same could not be said for the intellectual environment within which Hutton produced his 'Theory' the century before. Only Hutton and a few friends believed in his theory of geology, and a whole science had to be painstakingly reconstructed before any substantial part of it made sense on its own.

By 1890 North American geology and geomorphology had matured and could compete on an equal footing with the best that the Old World had to offer, as the Annual Reports of the USGS from 1880 onwards amply testified. By 1890 the subject had its own name, and the next Chapter will consider the effect that the Davisian Geographical Cycle has had on twentieth-century geomorphology.

Part IV

A SYSTEM AND ITS FEEDBACK: ANOTHER CENTURY OF DEBATE

Denudation and deposition seem to me clearly
incompetent to perpetuate their own cycles.

(T.C. Chamberlin, 1909)

Showing Alluvial Terraces of Soft Material Rapidly Eroded hy a River, which is Constructing what in Time
will be a yet Lower Terrace.

Chapter Nine

THE GEOGRAPHICAL CYCLE AND ITS TWENTIETH-CENTURY INFLUENCES

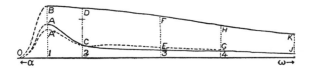

An imaginary landform.   (W.M. Davis, 1899a)
(L.C. King, 1953)

THE GEOGRAPHICAL CYCLE OF W.M.DAVIS (1850-1934)

No single theory in geomorphology has been so <u>influential</u>, and so widely <u>misunderstood</u>, as the Geographical Cycle. It was first formally enunciated by Davis in 1889, but the central ideas had already been hinted at in previous publications (Davis 1884, 1885). I suggested at the end of the last Chapter that the Cycle, as I shall term it for brevity, contained in essence nothing new. It used the erosional experience gained in the American West together with the views of Jukes (1862) on the regional development of drainage systems, and a new terminology, to establish a 'complete cycle of river life' (Davis 1889). The Cycle as announced in 1889 was used in a lengthy analysis of the rivers and valleys of Pennsylvania, a now classic area for the interpretation of landscape under humid temperate processes. The following excerpts are taken from Davis (1889) and illustrate 'The Complete Cycle of River Life: Youth, Adolescence, Maturity, and Old Age.' However, they do not purport to be a self-sufficient statement of the Cycle.

> A river that is established on a new land may be called an original river.. It must at first be of the kind known as a consequent river, for it has no ancestor from which to be derived...Once established, an original river advances through its long life, manifesting certain peculiarities of youth, maturity, and old age, by which its successive stages of growth may be recognized without much difficulty.. In its infancy the river drains its basin imperfectly, for it is then embarrassed by the original inequalities of the surface, and lakes collect in all the depressions..As the river becomes adolescent, its channels are deepened and all the larger ones descend close to base-level..With the deepening of the channels, there comes an increase in the number of gulleys on the slopes of the channel..With

their continued development the maturity of the system is reached; it is marked by an almost complete acquisition of the original constructional surface by erosion under the guidance of the streams. The lakes of the original imperfections have long since disappeared; the waterfalls of adolescence have been worn back..In the later and quieter old age of a river system, the waste of the land is yielded more slowly by reason of the diminishing slopes of the valley sides.

The most complete statement of the Cycle in English occurs in Davis (1899a) and various essays which treat of individual elements, or respond to criticisms are Davis (1896, 1899b, 1902, 1905a, 1905b, 1922). Because most of these are readily available in the 1954 reprint of 'Geographical Essays' (Johnson ed 1909) I shall assume that the reader has access to them. Recently a set of notes taken during Davis' lectures in 1926/7 has been published (King and Schumm 1980). These show that quite considerable modifications were envisaged for the Cycle as it applied to arid regions, quite late in Davis' life, in the light of his increased experience in, and west of, the Rockies. A factor of importance in the spread of the Cycle was the persuasive eloquence of his sketches and diagrams (Figure 12).

Figure 12. *Landscape sketches by W.M. Davis. Front ranges of the Rockies south of Denver showing monadnocks (above) and incision (below) (from Davis 1912).*

All of Davis' essays are so long that they defy a ready summary, and even his complete bibliography occupies some twenty pages! A comprehensive summary of his life and work is available in Chorley et al. (1973), so that here I shall try to concentrate on the impact that the Cycle had, criticisms of it, and alternatives proposed to it.

Whatever may be said for and against Davis' version of the Cycle the time was ripe at the end of the nineteenth century for the formulation and adoption of a broadly theoretical scheme integrating the available facts in the light of known landscapes. Gilbert (1877) noted the end results of prolonged erosion and De la Noe and De Margerie (1888), deriving much evidence from the Western Surveys, noted that a relative age might be assigned to landscapes on the basis of their slope form. Albrecht Penck (1886/7), independently of Davis, came to the idea of the peneplain 'Endzeit der Erosion,' although not the word, and Hettner (1921, 1928, trans. 1972), for long a bitter critic of Davis, remarked somewhat petulantly 'I developed the idea myself even before Davis did.' Newberry (1876) had referred to 'cycles on cycles' of denudation in the Colorado region, and so if it had not been Davis, it would have been someone else, because the very receptivity of the academic world to the Cycle indicates that it fulfilled a widespread need.

A new terminology for landscape analysis

Davis' most important contribution was the introduction of new terminology. As he himself noted:

> The introduction of a definite name for a thing previously talked about only in general has promoted its consideration: witness the name antecedent..it became popular when Powell named it (Davis 1899b).

Naturally, these introductions encountered a mixed reception and the majority of the several hundred new terms introduced by Davis and his followers are now defunct, but a substantial number live on in general use. Semantic complaints about the use of the new terms abounded, including the term 'Cycle' itself, and the deductive end state, the 'peneplain.' Subsequently the term 'grade,' applied to rivers, created a separate thread of debate which was eventually highly critical of Davis, and although he made it clear that both the idea and the word came from Gilbert (1877), and that it was merely a synonym for the 'profile of equilibrium' borrowed from the French engineers (Davis 1899a), he changed its exact meaning to suit his own ends.

In a discussion at the Royal Geographical Society of a paper by Marr (1896) on the English Lake District, the discussants refer to the advantages of 'a precise terminology' for 'the elucidation of practical problems,' although it was not thought useful for popular discussion! Reference was made to both Davis and de Lapparent, indicating that the French had already adopted the new ideas. However, at a later date Bonney (1912) could not see that 'consequent' and 'subsequent' held any advantages over the terms used by Jukes in 1862: 'transverse' and 'longitudinal.'

The Germans, in particular, had great difficulty in applying the terms implicit of age: infancy, youth, maturity, old age, because they saw that in a strict sense different parts of the cross-profile of a valley had different ages, and in addition it was their habit to refer the age of a valley line to the period of its origin. In contrast, Davisian terminology might give different ages to different parts of the same stream. Bowman (1926) referred to the 'inexplicably persistent misunderstanding of stage to mean age' whereas stage was meant merely to imply the amount of work done in the landscape. However, in retrospect it is easy to see how the application of a temporal scheme, to a process that was necessarily nonlinear in time, was bound to cause endless confusion in minds less finely tuned than that of Davis. The anthropocentric approach was not new with Davis, for it goes back at least to Du Buat. It was peculiarly appropriate at the end of the nineteenth century in the immediate aftermath of the Darwinian furore, and certainly helped popularize the Cycle, especially once it was entrenched in popular texbooks (Scott 1897, Russell 1898), from which even now it is not eradicated. In the late twentieth century it is something of an embarrassment, like a distant elderly relative at a teenager's party.

However, in supplying a terminology Davis ensured that geomorphology had its own language and could develop relatively independently of geology. Most importantly the new terminology was essentially morphological rather than structural and helped correct a tendency in some early geomorphological textbooks towards the sterile structural classifications of Powell: compare Geikie (1898) which is strongly structural, with Marr (1900) which is rather more Davisian. Moreover, a terminology is self-selective, users can take what they need without affecting the remainder.

Acceptance and spread of the Cycle abroad

Beckinsale (1975) has provided a convenient review of the spread of the Davisian Cycle upon which this section is based. The Cycle was received in France with almost universal acclaim and was adopted into the writings of de Margerie, de Lapparent, de Martonne and Baulig with a vigour that maintained momentum until the 1950s. Davis lectured at the Sorbonne for a year in 1911/12 and the explanatory and diagrammatic elegance of the scheme attracted the French. French influence on behalf of the Cycle was subsequently persuasive in Poland, Yugoslavia, Romania and Italy.

In Germany, however, the reception was much more mixed. Davis lectured in Berlin during 1909 and it is curious that his most complete statement of his system, with all its complications, was published in German in 1912, but has never appeared in English. The initial reception to the Cycle, based on its earlier appearance in English, was relatively favourable and Albrecht Penck wrote sympathetically about the new terminology in 1908. On the other hand, Passarge and Hettner resisted, and Hettner's book 'The Surface Features of the Earth' (1921, 1928, trs. 1972) was written specifically to warn his colleagues against the errors he perceived in the Davisian scheme, while urging them to re-evaluate their approach to the subject. This provides some indication of the popular appeal that the Cycle had in Germany. Albrecht Penck subsequently rejected the Cycle and its terminology in favour of a climatic approach to landforms and a more realistic recognition of tectonism. Penck's son Walther developed his own very original views on landscape evolution which will be reviewed below, and his early death in 1923 left his father as an apologist and propagandist on his behalf.

German influence in Europe was dominant in Bulgaria, Czechoslovakia, Greece and Russia, so that the Cycle was not strongly propagated in those countries, although it could be read in its German version (Davis 1912, 1924).

In the Great Britain the Cycle was welcomed since even with its 'complications' it was remarkably sympathetic to the needs of British topographic analysis. The fluvial adaptions to structure, the rejuvenations of baselevel, the abundant independent structural blocks with an apparently simple tectonic history, and the humid temperate climate all seemed to provide an ideal testing ground for the Cycle. Mackinder, Herbertson, Marr and Mill welcomed it and used it, either for strictly geomorphic purposes, or as the descriptive basis for the newly-developing subject of Geography. After World War I, Wooldridge and Linton used the Cycle, in association with the eustatic ideas of Baulig (1935), in their studies of southern England which culminated in 'Structure, Surface and Drainage in South-East England' (Wooldridge and Linton 1939). It is difficult, therefore, to understand Hettner's remark; that 'the sober English have largely discarded it.' He perhaps mistook a low level of research activity, and dynamic modifications in applying it, for a rejection of the single Cycle scheme. In fact, the personality of Wooldridge ensured that the modified Cycle (Wooldridge and Morgan 1937) remained a central force in British geomorphology until the 1960s.

The British Empire provided another haven for the Cycle until the 1950s. Charles Cotton in New Zealand remained an ardent, though flexible, Davisian from 1912 until his death in 1970, and his textbooks were widely used in Britain. New Zealand provided an interesting area for topographic analysis since it was volcanically and tectonically active, and still harboured glaciers. Since eustatic changes were global it was therefore an ideal region for the illustration of Davisian analysis under the impact of virtually all the major 'accidents' and 'interruptions' that Davis had envisaged. Lester King was a student of Cotton and started his career in South Africa during World War II. He soon came to doubt Davisian details and modified the cycle in the light of the semi-arid processes at work in South Africa, much as Davis himself had done in his later years, after his experiences in the American South West. Despite King's rejection of the Cycle as described, and his substitution of the pediplain for the peneplain, his work remained completely cyclic in concept (King 1953).

Australia provided one of the first stringent criticisms of the universal applicability of the Davisian analysis from the pen of Marks (1913) who, from the humid tropics of Queensland, pointed out the relative importance of chemical, as opposed to mechanical, weathering. Studies of the Burdekin river forced him to the conclude that the topography there:

> owed its differentiations from the surrounding hills only to the different weathering powers of the component rocks, and not in any way to a difference in age of denudation (Marks 1913, p98).

The complaint fell on deaf ears however, and in the hands of Edgeworth David and Griffith Taylor the approach was strictly Davisian within the University at Sydney. Finally, Japan adopted the Davisian scheme under the influence of Yamasaki, who had travelled with Davis in Europe, and Tsujimura, both at Tokyo. Davis' elementary textbooks on physical geography were translated into Japanese in 1930.

The domestic impact of the Cycle

One would expect that in the United States, where the architect of the theory resided, its effect would be overwhelming. Certainly the impact of Davis was enormous, his Chair at Harvard, and his presidency of both the Geological Society of America, and the Association of American Geographers ensured that he was known personally to all the principal personalities on the continent. Yet there remained an undercurrent of suspicion about the Cycle from its earliest days. It is difficult to articulate the objections in a coherent way. Early objections were largely on semantic issues of terminology (Tarr 1898, Davis 1899b, 1902), and many of these were raised privately in correspondence, but answered publicly by Davis in his printed works without a clear attribution to their source. The term 'peneplain,' invented by Davis, was objected to on the basis that examples close to base level could not be found, or that none existed. As Bishop (1980) has shown, Davis answered these, in part, by qualifying the character of the peneplain with other terms taken from the Cycle. In this way he rendered the term unassailable or 'unfalsifiable' and to this, according to Bishop, can be attributed the longevity of the Cycle and its terminology. Davis would have agreed with King (1953) that it was 'an imaginary landform,' taking as a compliment for deductive thinking a slight to sloppy theorizing.

Notable physiographers approximately contemporary with Davis were Bowman, Salisbury, Atwood and Fenneman, and to a greater or a lesser degree they accepted Davisian teaching although Salisbury in later editions of his 'Physiography' (1924) considerably reduced the space devoted to the life history of rivers. Although the Cycle is an ideal teaching device, it is very simple indeed in the single cycle form and requires little space and just a few diagrams. However, Davis (1889) introduced it formally as a research device, and in this guise it could be, and was, expanded imaginatively to cope with any contingency. But, in this form, it is not easily summarized for an introductory text. Nevertheless, there can be few papers on regional physiography published in the United States between 1890 and 1940 which can claim to make no use whatsoever of Davisian notions.

Apart from the 'accidents' of volcanism and glaciation, the main 'interruption' of the Cycle was thought to be a relative change in base level, implying either vertical movements of the crust, with some degree of tilting, or eustatic shifts of sea level. Any such change (Davis preferred epeirogeny, the British and the French eustasy), initiated a new cycle and in the coastal depositional system this would lead to a new phase in the stratigraphic record. Because the Cycle was so well-adapted to prevailing diastrophic thought in geology its adoption was natural to geologists whose main professional concern was not with the present surface, but with understanding buried unconformities and stratigraphic sequences related to erosional episodes on the land. Significantly, therefore, the tide only began to turn against Davis when increasing attention was paid to the physical processes of the surface during and after the Second World War, by geomorphologists such as Bryan (1940), Horton (1945) and Strahler (1950a), and when the prevailing diastrophic model also began to crumble (p117).

Extensions and adaptions of the cyclic idea

The fundamental idea that landforms, under a stable base level, undergo a sequential series of stages was easily adapted to landform systems to which the original fluvial Cycle was inapplicable. Davis suggested both the coastal and the glacial cycles (1899a) and his student Gulliver (1899) described a coastal cycle that was later refined by D.W.Johnson (1919). In his youth Davis (1882) had doubted the efficacy of glacial erosion but he confessed his change of mind (Davis 1900) and described 'The Sculpture of Mountains by Glaciers' (Davis 1906). W.H.Hobbs (1910, 1911) developed his own version in rather more cautious terms 'since in our use the cycle is measured in climatic changes rather than the attainment of certain denudation effects within glaciated valleys.' Nevertheless, Hobbs did attach age terms to characteristic sequential relief elsewhere in his paper.

The Cycle was adapted to arid lands in a paper in which Davis (1905) acknowledged his 'special indebtedness to Passarge' (1904), for his descriptions of the Kalahari. A more extreme version of the arid cycle involving only aeolian action was published by Keyes (1912). However, by the 1920s Davis' lecture notes make it clear that he had modified his understanding of the mechanisms at work in arid landscapes considerably and he drew attention to the work of McGee (1897), Lawson (1915) and Bryan (1922) who introduced the word 'pediment' for rock-cut surfaces around mountains. After World War II, Lester King developed the idea of the semi-arid cycle and the merging of individual pediments to create pediplains. His major contribution was to extend this hypothesis to the major plainlands of the world, arguing that before the development of turf most land surfaces behaved as present day pediments. He connected the genesis of these major cycles with world-wide tectonic episodes and the theory of Continental Drift (1953, 1962).

Karst regions attracted the attention of Sawicki (1909) in Poland, Beede (1911) in the States, and Cvijic in Yugoslavia. Cvijic published an excellent mongraph on Karst in 1893 under the direction of Albrecht Penck. His cyclic interpretation, published in France during World War I, was popularized for the English speaking world by Sanders (1921). A lost manuscript by Cvijic was re-discovered and published in French translation by de Martonne (Cvijic 1960). In the twilight of his life Davis wrote an influential paper on the origin of limestone caverns (Davis 1930) in which he proposed a two-cycle origin for cavern systems. The paper rejuvenated thinking on karst geomorphology, producing, for example, important papers by Swinnerton (1932) and Bretz (1942).

The first application of cyclic ideas to humid tropical landscapes was apparently made by Wayland (1933) who described deep weathering below a planated surface in Uganda which was then easily stripped off upon renewed uplift. He termed the resulting surface an 'etchplain.' Later Cotton (1942, 1961) summarized subsequent work on a somewhat similar process that he called 'savannah planation.' At the other end of the climatic spectrum Peltier (1950) identified nine morphogenetic regions, each of which was locally 'normal,' and then outlined a periglacial cycle to cope with areas strongly affected by Quaternary glaciation.

Even the humid Cycle itself was not immune to radical modification. Johnson (1931) suggested that later stages in the Cycle are dominated by lateral corrasion from river meandering, from which Crickmay (1933, 1974, 1975) deduced the eventual creation of a 'panplane.' More fundamentally, Fenneman (1936) added the element of 'non-cyclic' erosion: an adjustment of surface to structure complementary to the idea of streams adjusting to structure.

ALTERNATIVES TO THE GEOGRAPHICAL CYCLE

The all-embracing nature of the Davisian imagination, and the Cycle was intended to be a complete system of geomorphology, left conceptual competitors little room to manoeuvre. However, alternative viewpoints did develop, either as national schools of thought, for example the French and German schools, or as rather isolated pieces of work independent or critical of Davisian concerns, for example developing research on surface processes and the original endogenetic views of Walther Penck. These I will discuss under several different headings.

Climatic geomorphology

The first formal paper stressing the importance of climate to physiographic development was by Albrecht Penck (1910) who was stimulated by Supan (1884) and Koeppen (1901) into seeing whether physiography, rather than biogeography and climatic statistics, could define climatic regions. His main interest was the reverse; did different climates determine distinctive assemblages of surficial processes? He identified the humid, nival and arid regimes, and within the humid regime noted the phraetic and frozen ground subdivisions. The recognition of periglacial processes was quite explicit: the high river flows of summer, the mobility of the active surface layer, and the general accord of the southern limit of frozen ground with the minus two degree Celsius isotherm.
  He recognized too the special case of karst, and that transitions existed between regimes, giving rise, for example, to the existence of surface crusts in semi-arid areas, but their absence in truly arid regions. As Budel (1948) pointed out much later, 'the Penckian system lacks a morphological basis;' that is it did not specify what forms would arise, it merely drew attention to the specialized surface processes under different hydrographical regimes. In this sense Penck deserves equal recognition with Gilbert as a founder of the modern school of process geomorphology. Subsequent workers, however, could not resist the temptation to subdivide the surface on the basis of increasingly refined climatic parameters, and naturally there was no shortage of small scale landforms to justify the decisions. De Martonne (1909), like Davis, discussed initially only glacial,

temperate and arid landscapes but he later remarked (1913) that 'new horizons open before those who embark on the study of more exotic regions where a poorly understood suite of processes prevails in extreme conditions of temperature and humidity.' His second scheme identified six 'topographical facies' and 'innumerable transitions between the different facies' with the complication 'that the landscape tends to preserve evidence of past climates which differ from the present.' However, De Martonne, and his pupil Birot, were both primarily Davisian and a strong French approach to climatic geomorphology did not develop until after World War II. Nevertheless, De Martonne had identified the two crucial notions of the climatic viewpoint: (1) the 'morphoclimatic' notion that particular forms spring from certain climatic regimes given sufficient time, and (2) that climatic types may not have been stable over geological time, and therefore complicate the interpretation of surface form. Subsequent work tended to reflect and substantiate these ideas and Budel (1948) stressed again the importance of 'paleo-landforms' to the interpretation of particular landscapes and originated the term 'morphogenetic region' to describe the response of landscape to unique sets of processes.

The French (Tricart and Cailleux 1965) usually preferred to use the term 'morphoclimatic' to express the notion that the surface responds to the total milieu of climate, soil and vegetation, as they condition surface hydrology and its impact on landforms. This implied a change of emphasis from those, such as Albrecht Penck and Thorbecke (1927a), who saw climate, in itself, as a sufficient basis for landscape classification. Passarge (1927) was an early adherent to this more general view and was the lone dissenter upon this point at an important conference at Dusseldorf in 1926 devoted to the morphogenetic point of view; and whose proceedings (Thorbecke 1927b) can be regarded as 'a founding document of climatic geomorphology' (Stoddart 1969).

In general, most of this work reflected a much greater concern with what Hettner (1928) termed 'minor landforms,' or attention to details, than was ever evident in Davisian analysis. The climatic viewpoint ought to have galvanised interest in the basic processes of weathering, but once more they were taken for granted and research concentrated on assessing the relative importance of weathering types in different climatic environments. An appreciation of the power of chemical weathering in the humid and seasonal tropics followed from the discovery of very deeply weathered profiles, up to hundreds of feet deep. The induration of certain mineral rich layers within the profile (termed a 'duricrust' by Woolnough (1930)) and the potential stripping of the weathered profile to reveal the uneven basal surface were two fruitful concepts for later research. In essence therefore, weathering was viewed in its regional setting and because both cyclic and climatic schools still adopted a cyclic timeframe, there was small incentive to carry out field or experimental work on processes whose general results were assumed to be understoood, and for which the basic climatological data was often meagre or non-existent.

Perhaps because of these factors there was much common ground, as is revealed in the title of Peltier's paper (1950) 'The Geographic Cycle in Periglacial Regions as it is Related to Climatic Geomorphology' which built a periglacial cycle upon fifty years of gradual progress in the description of periglacial terrain, including cyclic suggestions (Matthes 1900, Andersson 1906, Lozinski 1912, Cairnes 1912, Bryan 1946, Troll 1948). It was possible to belong to both camps so that on one level the differences are merely those between 'lumpers' and 'splitters' in the jargon of taxonomy. Laying aside karstic, glacial and coastal cycles as 'special' Davis himself dealt only with humid and arid regions, although late in life he gave explicit recognition to the semi-arid case for the American South-West and realized that he had over-emphasized the work of wind in arid regions. At the same time the French and Germans recognized at least four or five morphologically significant regions. Wayland (1934), Cotton (1942), Peltier (1950) and King (1953) all added to the cyclic stock so that there was, to an extent, a convergence of opinion if not in emphasis.

A recent book of readings on climatic geomorphology (Derbyshire ed. 1973) included an edited version of the Cycle because of its importance in understanding climatic geomorphology in this century. The editor also observed that 'traditional climatic geomorphology .. glossed over this paucity of knowledge of fundamentals' and 'proceeded, like Davis' work, to premature generalizations on the basis of quite vague ideas about the underlying process relations.' The climatic approach offered, therefore, a different perspective; one based on smaller scale features, an explicit recognition of local detail, and the importance of geologically recent shifts in climate. Stoddart (1969), in a comprehensive review of the rationale of climatic geomorphology, concluded that 'it was rooted in classical cyclic geomorphology' and has eventually returned 'to an equally historical position, indistinguishable except in flexibility from classical geomorphology.' However, there are significant differences in approach other than those which arise from an emphasis on climatically or environmentally controlled suites of processes.

One difficulty in assessing the work of the German school of climatic geomorphology has been the difficulty of ready access to a literature that is rarely taught as a second language in the Anglo-American world. As a result much of the transfer of ideas has been indirect via the French school led by Cailleux and Tricart, although the evangelism of Cotton (1958, 1962) is a notable exception. The leading German theoretician has been Julius Budel who began his field work in the early 1930s, and who has remained active throughout five decades. A comprehensive volume summarizing his life's work and thinking was published in 1977, and in English translation in 1982. The remainder of this section will try to give the flavour of the climatic paradigm for geomorphology mainly as related by Budel, and which, as the translators' preface remarks, is at least representative of German geomorphology. The research recounted by Budel (1982) mainly relates to the period after World War II, when the Davis model was being progressively abandonned in the Anglo-American world to be replaced with a growing interest in small scale, short term processes. The details of these shifts in interest over the last four decades will be found in the next two Chapters. Most the ideas outlined below appeared in earlier papers by Budel (1948, 1957, 1963), but the recent volume is a definitive summary.

With Budel, his translators' remark, it is but a small exaggeration to speak of 'a distinctive Budel language,' for like Davis before him, he found it necessary to develop an entire terminology to encompass a quite different conceptual view of topography.

Budel's temporal stage is that of cyclic time, millions of years, and he calls the period since the late-Cretaceous the 'geomorphic era': the one during which, and with only limited exceptions, the world's relief has been created. This in itself sets his interests apart from mainstream work in Britain and North America in the post-War period. His attitude to Anglo-American work is stated thus:

> When individual analyses or mathematical models are applied without recognizing their place in the multi-levelled integrational system of geomorphology, no matter how rigorous or logical the attempt in itself, unnatural models and inappropriate simplifications may result. For this reason I refuse to consider such randomly performed analyses as building blocks for an exact geomorphology.

Similarly he believes that 'the complex interactions of all the component processes' which leads 'to the creation of relief...cannot be replaced by the cleverest of simulations.' His emphasis throughout is upon the interactions of many complex elements in the physical landscape to create what he terms 'relief-forming mechanisms.' The basis of this analysis is climatic geomorphology which reveals the interacting complexes which form relief. The contemporary operation of these mechanisms, and the identification of the characteristic relief to which they lead is termed 'active geomorphology.' Climatic influences form the best basis for a system of geomorphology because 'the exogenic effect of climate follows strict rules in its distribution from the pole to the equator.'

The diastrophic framework of Budel's thinking differs radically from that which is conventional in Anglo-America. Although he reviews and accepts plate tectonics and modern endogenetic concepts he sees, from the point of view of geomorphology, that 'endogenous forces, as any geological map will show, are distributed randomly over the entire earth' with the implication that they are to be taken as found, rather than integrated logically into the sequential development of landforms. In this attitude he sides with Davis insofar as endogenous inputs are regarded as external, time independent controls, but he differs from him in paying no regard at all to base level as a significant concept controlling the erosion of regional landscapes. Needless to say he differs from Walther Penck (see below), although he recognizes the 'primarrumpf' concept as valuable.

Louis (1957) and Mabbutt (1961) also have addressed the problem of base level control and make it clear that in etchplain landscapes the basal surface of weathering may be quite independent of base level; local or global. However, most of the criticism of the base level concept is addressed to the unrealistic single cycle case rather than its actual role in a multi-cyclic landscape.

The temporal perspective is organized in Budel's work by making a virtue of the necessity that most places on the earth's surface, especially in mid to high latitudes, have undergone radical climatic change since the late Cretaceous. Thus most landscapes carry the imprint, and it is more than a palimpsest, of former relief-forming mechanisms. The interpretation of successive 'generations' of relief formation is deduced by 'chrono-geographic comparisons:' by comparing the results of contemporary active geomorphology in regions of similar physical geography, with the relict forms awaiting interpretation. In this way morphoclimatic geomorphology illuminates the historical record and the technique, which is not without its dangers, is a restricted type of the ergodic hypothesis (see Chapter Eleven).

The central task of Budel's geomorphology is therefore seen to be climato-genetic geomorphology: 'recognizing relief generations, separating their remains, and unraveling their developmental environment.' The effect of subsequent processes on relic relief forms is termed 'synactive geomorphology' an enterprize which must determine the extent to which the surface has been destroyed, transformed or perpetuated. The application of these ideas to particular landscapes, especially the morphoclimatic conclusions of active geomorphology, produces some surprising conclusions. In Central Europe the topography is seen to be the accumulated results of four relief generations beginning with the Tertiary etchplain stairway systems, now largely stripped of their deep weathering profiles, but which began as the double planation surface systems now forming in the 'peritropics' (for example eastern Africa, India, and northern Australia). There followed in the late Pliocene a period of valley cutting which fixed the main river systems of Europe, although Budel admits that there are difficulties in identifying the morphodynamic processes attendant on this short (two million year) phase.

The third period relates to the Pleistocene Ice Age and was morphologically transgressive to the pre-existing relief in at least two ways. In areas of active ice sheets erosion and deposition produced a diverse, though relatively well-understood landscape, whereas in the periglacial zone peripheral or subsequent to Pleistocene Ice, two processes dominated: excessive valley cutting and active solifluction on slopes. The most interesting aspects of the interpretation of glacial relief is that the bulk of it is thought to still carry the imprint of the Tertiary etchplains. Trough shoulders and longitudinal valley steps are regarded as remnants of etchplain stairways, and roche moutonees are merely inequalities on the basal weathering surface, stripped of its weathering products, and smoothed by glaciers. Thus a form of glacial protectionism and selective linear erosion is deduced, a result not inconsistent with some Anglo-American work in Arctic regions.

The fourth major relief generation is that of the Holocene for which the morphoclimatic processes are classified as belonging to the 'ectropic' zone of retarded valley-cutting where soils and vegetation stabilize the relict relief.

Some of these conclusions have appeared sporadically in Anglo-American work, possibly independently, possibly by transfer. Linton (1955) developed a theory of tor formation drawing on the notion of tropical deep weathering and subsequent stripping to reveal elements of the basal weathering surface. The idea was previously developed by Mortensen (1932) and the distribution of castellated tors in Britain was also discussed by Louis (1934) although his evidence was very incomplete. The etchplain concept, originated by Wayland (1920, 1933) and elaborated by Budel for the seasonal tropics has found a reasonable acceptance in the English speaking world. The selectiveness of glacial erosion has been noted already, and the general problem of very active valley cutting during parts of the Pleistocene has been extensively investigated by Dury (see Chapters Ten and Eleven) in work which is not referenced at all by Budel.

The work of the German school of climatic geomorphology can be viewed as a logical development of Davisian thinking, provided the latter is taken in its least restrictive sense. It did not get well under way until after World War II, by which time the Davis model was in decline, and it has pursued a relatively independent course although its main lines of thought have filtered through to Anglo-American thinking or have evolved independently, especially with respect to low latitude landforms. Even with more thorough transfer it is unlikely that it would have had a major influence on work being conducted at different spatio-temporal scales. Its idiographic slant would not have appealed to geomorphologists from a geographical background that was itself undergoing a revolt from regionalism, nor would its cavalier attitude to diastrophism and structure have endeared it those trained in geology departments. Complicating problems are that there is a marked disagreement about the global identification and distribution of morphoclimatic zones (Tricart and Cailleux 1965, Stoddart 1969), and it is not at all clear whether morphogentic deductions can be drawn unambiguously from the existing physical milieu. Anglo-American papers usually take the graphs in Peltier (1950) as the basis of morphogenetic discussions, and significantly these isolate processes rather than delineate spatial regions (Figure 13).

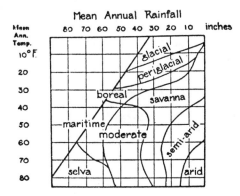

Figure 13. *Peltier's (1950) diagram of morphogenetic regions.*

The sequential series of landforms that climatic geomorphology discovers bear only a superficial resemblance to the Davisian sequence, not least because of the different treatment of baselevel, yet in the long term some compromise must be reached between the climatic school, the modern process school and an eventual mutual convergence of interest on what Budel rightly calls the 'geomorphic era.'

Pre-War work on surficial processes

Conventional thinking regards modern research on surface processes as post-War, but in fact a number of significant threads of activity were underway during the 1930s even though these are poorly represented in the spate of text books that ended that decade (Wooldridge and Morgan 1937, Lobeck 1939, Von Engeln 1942).

In the central field of fluvial systems Horton (1932, 1938) was publishing work that was later summarized in his famous paper (1945) on the hydro-physical approach to drainage basins. Horton (1875-1945) was a consulting hydraulic engineer with his own laboratory and his publications in the thirties coincided with growing public concern about soil erosion, both by wind and water, and the preservation of agricultural land. These concerns climaxed in 1935 with the establishment of the Soil Conservation Service under the leadership of Bennett who had campaigned long and hard towards such an end, and whose book on 'Soil Conservation' (1939) rapidly became a classic in the field. Horton's papers concerned soil erosion by overland flow on experimental plots, but they attracted no attention from geomorphologists until after the War.

At a scale closer to that of the cycle Sharp (1938) published an influential classification of types of mass-wasting based on the proportions of water, rock and ice involved in the movement, and on the character of the movement itself. Behre (1933) made an early study of talus slopes relating measured surface angles to the motion of the fragments down the slope, although his conclusions were disputed by Bryan (1934). Studies in weathering lagged behind inferences about the origin of landforms in the nineteenth century, and the story was repeated in the first half of the twentieth century, probably because of a continued disparity in the spatial scales of interest. Merrill's (1897) review of weathering was not materially superseded until Reiche's survey (1945) and the most significant papers were probably those by Blackwelder (1933) and Griggs (1936) which showed the importance of water cooling to the fracturing of heated rocks, a technique long known to quarry men and reported by Stanley (1876). One important book which appeared during the 1920s was Glinka's book (trs. 1927) encapsulating the Russian work on the formation of the great soil groups of the world. The global relation to climatic factors, and the notion of zonal and azonal classification and their dependence on the sum of environmental influences was a significant stimulus to the climatic school of geomorphology.

The focus of fluvial activity is the river. Davis, late in his career, included elementary hydraulic information in his lecture notes (King and Schumm 1980), but such information never affected cyclic thinking. Eventually a beginning was made towards relating the properties of flow to the morphology of river channels by Leighly (1932, 1934) who directed his attention to the importance of turbulence in effecting transportation of the suspended load. The 1932 paper appears to have been overlooked in subsequent

discussions of process and form in river channels: Mackin
(1948), in his discussion of river grade, cites only the less
crucial 1934 paper. In it Leighly connects the radius of
curvature of the meander bend and the rate of change of
hydraulic slope along and across the water surface with its
propensity to scour, deposit and change the channel form.
Many statements in the paper anticipate post-War research. He
deduced that the interaction of stream and bed is very
different between consolidated and unconsolidated bed
material and concluded that the:

> assumption of downward cutting as characteristic of all
> streams, in their development toward equilibrium, is
> unjustified. In unconsolidated rock, the equilibrium may
> be attained by a widening and shallowing of the stream.
> Any slope to be encountered in unconsolidated material
> may be a slope of equilibrium for a stream flowing over
> it, if the the width and depth of the water channel are
> properly adjusted: which adjustment appears to be
> attained automatically by the stream itself. The cutting
> of an ever deeper gorge, until the hydraulic slope of a
> stream of a given discharge becomes adjusted to the
> material carried, and the walls of the gorge are perhaps
> reduced to gentle slopes through denudation, is a
> function of streams whose beds consist of material which
> acts as "consolidated" at those river stages which
> determine the form of the channel. Transverse profile
> and the plan of the channel are determined by the nature
> of the reaction of bed material to turbulent flow.

Thus in a few sentences he related hydraulic behaviour to the
excavation of valleys, a crucial topic that is still scarcely
understood. In the second paper he analysed the distribution
of turbulence in the cross section and noted that
sedimentation in rivers was primarily lateral, thereby
affecting channel morphology which, he remarked, changed most
rapidly 'during periods of fluctuating discharge.' The idea
was applied to the infill of oxbow ends, to point bar
creation, to levees and, he speculated, to the shifting
channels of wide, shallow rivers. When in 1940 Leighly
remarked that 'Davis' great mistake was the assumption that
we know the processes involved in the development of
landforms. We don't; and until we do we shall be ignorant of
the general course of their development,' he spoke as one who
had actively tried to rectify matters.
    In his 1934 paper Leighly noted the concurrent
appearance of a paper by Hjulstrom (1932) on the
morphological activity of rivers, and one by O'Brien (1933)
also examining the relation of suspended sediment to
turbulence. A later paper by Hjulstrom (1935) is more
frequently cited since it provided a convenient graph showing

the erosion, transportation and sedimentation velocities required for particle sizes usually encountered in rivers. Additional studies on sediment transport in natural rivers were made by Rubey (1933, 1938).

Pre-War papers on coastal processes are reviewed in Chapter Ten since they led directly into wartime work of immediate military relevance. Military interests were behind the researches of Bagnold (1933, 1935, 1937) in the North African desert. Bagnold noted the lack of connection between the description of large scale dune fields and the basic mechanisms of interaction involving a sand grain and the wind. He applied physical theory, wind tunnel experimentation and judicious field work to the elucidation of sand movement and dune morphology. His collected results appeared in 1941 as a book which was quickly recognized as a classic.

Whatever lines of research these early papers on process may have led to, individually, or as a significant new trend in geomorphology, was interrupted by the War. After the War, similar but not identical threads can be identified. However, the immediate impetus had been lost and almost two decades were wasted regaining lost ground, and the prescient papers of the thirties have lain largely forgotten.

Non-cyclic views of landscape

To adopt a non-cyclic view of the landscape in the early twentieth century was to ignore millenia of intuition which had deduced that continued denudation would reduce uplands to plainlands. It was also to strain the imagination about the earth's diastrophic behaviour, as it was then understood.

At the spatial and temporal scale of the Cycle the only relevant work is that of Ashley (1931, 1935) and Fenneman (1936), the true significance of which may not have been realized. Ashley emphasized the late Tertiary character of the scenery, and both he and Fenneman were concerned with the multiplicity of erosional levels, all termed 'peneplains' in conventional discussions, that were being identified by their contemporaries in the Appalachians. This concern was echoed a little later by Rich (1938), but his solution was different. Fenneman thought the matter so important that he made it the subject of his Presidential address to the Geological Society of America. He observed that:

> the assumption that a series of peneplains reflects changes of base-level throughout the area concerned carries with it the assumption of an accelerated succession of diastrophic events. A very long series would seem to involve serious issues in which geophysicists may well ask to be heard. If a dozen or a score of base-levels be evidenced, the probability becomes very great that most of them were local.

He classified geologists as 'structure minded' or 'cycle
minded' in terms of geomorphology and pointed out that the
peneplain connoted 'a mode of origin quite as much as a
form.' The Cycle he noted was 'not a physical process but a
philosophical conception. It contemplates erosion in one of
its aspects: that of changing form. But erosion does not
always and everywhere present this aspect.' He went on to
define non-cyclic erosion as 'a slow wasting of a surface
without change of characteristic form:'

> So constant is our association of valleys with erosion
> that it is difficult to think of the straight, horizon-
> tal Appalachian crests as being lowered scores or even
> hundreds of feet and yet looking the same after as
> before.

He then suggested that the height of the crests should be
correlated with the character of the rock (cf Marks 1913) and
the breadth of the outcrop because of their potentially
differential response. It would then be possible to ask 'how
many baselevels must be assumed?..Perhaps three would be
enough, or two, the last and extremest suggestion is one.'
These quotations unmistakably foreshadow the non-cyclic
dynamic equilibrium theory developed much later by Hack
(1960). Fenneman foresaw that in the right diastrophic
framework 'there is no theoretical limit to the amount of
uplift and erosion that a maturely dissected plateau may
undergo without change of characteristic form.' He applied
these ideas mainly with the intent of reducing the
diastrophic burden that a multiplicity of surfaces created.
He thought that correlation corrections of the sort he
suggested should be made since 'to correlate without them
should be classified by law as a dangerous occupation.' He
followed these precepts himself since in his volume on the
'Physiography of Eastern United States' (1938) he remarked
that he would outline only major cycles. It is not
insignificant that his paper appeared at a time when cracks
were appearing in the prevailing theory of diastrophism
(Chorley 1963), and when the concept of isostatic balance in
the crust was slowly gaining ground (Lawson 1936) and it is
interesting that some of the strongest proponents of isostasy
have been geomorphologists cognizant that surface processes
can, given enough time, seriously disturb the isostatic
balance of a crustal column (Dutton 1889, Gilbert 1890,
Barrell 1919, Bowman 1926, Lawson 1936, 1948).
There is little sign that Fenneman's paper had any
significant effect on cyclic thought. He had not meant it as
an attack on the Cycle itself, merely as an important
modification. He remarked prophetically in conclusion that
'it matters not whether the terminology here used be liked or
accepted. It matters much that the facts be recognized.'

The Penck/Davis controversy

Walther Penck (1888-1923) was the son of Albrecht Penck, the famous German geomorphologist. He was determined to prove himself in a field other than his father's and so came to geomorphological work only after the World War I. He died of cancer at the age of thirty five and left an unfinished manuscript that was published by his father in 1924. By common agreement its German was tortuous and a published English translation only became available in 1953, although manuscript versions of parts of it existed in North America before World War II. The book is entitled 'Morphological Analysis of Landforms' and references to it will be to given as MA, with page numbers from the English edition (1953, trs. Czech and Boswell). Existing correspondence between the Pencks and Davis is reproduced in Chorley et al. (1973).

The controversy between Penck and Davis, at least as it entered the geomorphological world through printed sources, concerned two issues. One was the imprint that endogenetic forces were thought to have on slope form, and the other was the nature of the retreat of slopes. Insofar as many geologists dismissed outright Penck's views on the first issue, most discussion centered on the second point. The conflict has often been thought to represent one between rival American and German schools of thought, whereas in fact many German scholars rejected W. Penck as much as they rejected Davis. A. Penck's views were primarily those of the climatic school, and must be distinguished from those of his son which flatly rejected climatic differentiation of landforms as significant. How Penck senior resolved these conflicts for himself is not clear. W.Penck's book, although complete in itself, was both unfinished and unrevised: in time he might have changed his mind on many points of detail.

Regrettably, Penck's work became best known in the English speaking world, not in its own right, but through reviews by Bowman (1926) and Davis (1932), both of whom mis-interpreted the German texts and led to the propagation of serious errors concerning Penck's thinking. These errors were not corrected until the post-War papers of Tuan (1958) and Simons (1962). In the meantime, a largely groundless controversy raged for thirty years. It is advisable to begin with Penck's own ideas (MA 1953).

Penck stressed above all that 'the earth's surface is a field of reaction between opposing forces' (MA 3) and by this he meant exogenetic forces and endogenetic forces. He thought too that 'forms of denudation' should relate to 'correlated deposits' elsewhere. On the slope he assumed that the erosion rate was proportional to slope angle, and he adopted what he called the differential method to analyse the development of slope forms,' reduced to a myriad individual units. In the limiting case he collapsed these individual units into a continuous curve, a point missed by many analysts who insisted on taking Penck's individual facets as real. He deduced that 'the most important law obeyed during the development of denudational forms is this principle of flattening' (MA 121); another crucial point that Davis inadvertently missed. The results of his analyses led Penck to believe that the concave slope form was 'the normal type of form.' It should be pointed out too that the four chapters following the Introduction provide an excellent review of tectonism, weathering, mass movement and denudation, preparatory to the theoretical chapters on slope evolution which usually attract attention.

On the large scale he was convinced that 'areas of similar endogenetic development have also, in all climatic zones, the same denudational forms' (MA 119) although he did qualify this with the remark that they form 'under different climates at different rates' (MA 120). This, and other similar remarks, clearly reject the German school of climatic geomorphology. But Penck was probably not thinking of a rigorous geometrical scheme for the eventual slope form. His differential analysis was merely a deductive method for arriving at more general conclusions. He specifically illustrated (his Plate VIII) the varying form of inselbergs in different climates, and he wrote that 'some details of the characteristic features..may be traced back to climatic influences' (MA 196).

Penck assumed a familiarity with the Cycle amongst his readers and he observed that Davis had been misunderstood by his followers! Davis, he realized, was aware of the potential complexity of tectonic processes, (MA 12) but Penck saw that both Germans and Americans, at the time he wrote, were using or criticizing the simplistic single version of the Cycle. While he agreed that the Davisians could be chastized for not instituting a world-wide study of processes he remarked that this failure did not negate the Cycle since they assumed no principle which had not been verified. He did not object, as his compatriots did, to the deductive nature of the Cycle; nor could he, since his own scheme was equally deductive, if more explicitly so. His main objection lay with its failure to examine, or develop, the complications of the endogenetic processes that could affect landscape throughout its sequential development. Hence, as a working framework, Penck was in substantial agreement with the Cycle. No wonder that Hettner (1928) disdainfully dismissed them together:

> Most of the objections on which Davis' theory is to be rejected can, with little modification, be levelled against Penck's as well.

Bowman (1926), in reviewing 'Morphological Analysis,' claimed that Penck was really dealing with the 'complications' of the

Cycle, although A. Penck (1926) denied this. However, he wrongly attributed to Penck the idea that the slope profile revealed in its concavity or convexity the rate of uplift, and this error, reinforced through a restatement by Davis (1932), lived on in American geomorphology to be denounced by Douglas Johnson (1940) as 'one of the most fantastic errors ever introduced into geomorphology.' The problem may have lain in Penck's German because, in general, it was Penck's proposition that landform assemblages reflected spatial and temporal patterns of uplift. This position is virtually identical to that which the Davisians would have to accept, logically, for a landscape subject to many partial cycles: precisely the point, and the worry, that Fenneman (1936) was later to articulate.

Davis, to judge by the correspondence in Chorley et al. (1973), admired Penck's book although he had objections in detail. However, his main criticism of Penck was directed at another posthumous publication, one analysing the piedmont benchlands of the southern Black Forest (Penck 1925). Penck's analysis claimed to show that smoothly accelerating uplift could produce a benched massif, whereas Davis was convinced that the pattern revealed sporadic uplift. Davis' review was remarkable in view of his advancing age, but in the course of it, he reiterated Bowman's error and added another; that:

> When valley deepening ceases, the degradational retreat of the valley side goes on parallel to itself and leaves a surface of less declivity (a valley floor) below it as it withdraws (Davis 1932).

This, in the very best light, is a gross simplification, but it can only be seen as such by someone with an intimate acquaintance of Penck's text and analysis. To exacerbate matters, Davis included a diagram illustrating the supposedly contrasting modes of slope retreat (Figure 14) which still appears in textbooks (Muller and Oberlander 1978), despite Tuan (1958) and Simons (1962).

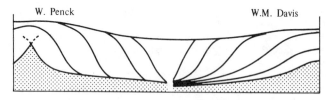

Figure 14. *The figure, from Davis (1932), which erroneously characterized the Penckian model of slope retreat as one of parallel retreat in contrast to Davis's slope flattening.*

At the Penck symposium held in 1939 by the Association of American Geographers most attention was directed at the two errors which Americans themselves had perpetrated, and the proceedings are largely nullified thereby. Indeed the comments show misunderstanding of Davis as well as Penck. Meyerhoff was invited on the mistaken premise that because he saw several partial cycles in New England, all in simultaneous development, he must be an adherent of Penck! (Meyerhoff 1976). Ironically, both Penck and Davis are most usefully seen as complementary to each other, and Davis, by 1930, accepted the parallel retreat of semi-arid slopes while thinking that Penck was preaching it for humid slopes!

Bowman (1926) thought that Penck's book could stimulate American geomorphology. It failed to do so in the way that it should have, and by the time the translation became available in 1953, slope studies were underway with the intent of discrediting Davis as much as Penck, although the supposed Penckian assertion was most usually tested.

CATASTROPHIC INTERLUDE: J.H.BRETZ AND THE CHANNELED SCABLAND

The early twentieth century prided itself for its acceptance of uniformity, and for having rejected all traces of nineteenth-century catastrophism. It had learned too, that field evidence was paramount. Thus, when Harlen Bretz (1923, 1925, 1928, 1932), mapping landforms in the Columbian plateau of eastern Washington, came to the conclusion that the only conceivable agent that could have caused them was flood water on a stupendous scale (he estimated discharges of 66 million cubic feet per second), the academic reaction was understandably skeptical, if not hostile. Baker (1978a) recounts the controversy. Specifically, Bretz noted huge gravel bars, some 30m high loess hills in the Cheney-Palouse region that were prominently streamlined, and the rugged scabland itself revealing anastomosing rock channels and basins cut in basalt by the stream-bed mechanics of the gigantic flows. The scale of the flows was so massive that what appear to be valleys were in fact the river channels of the flood: in places up to 5km wide and 100m deep. Bretz found the source of the flood in the abrupt drainage of Glacial Lake Missoula, when an ice dam failed; an analysis later confirmed by Pardee (1942) who described ripple marks 15m high resulting from what is now called the Spokane Flood.

During the 1930s considerable effort was directed to proving that the landform assemblages were caused by floating ice in water ponded by ice jams (Allison 1933), by lake fills, glacial outwash, and subsequent stream dissection (Flint 1938), or by a complicated series of glacial advances and retreats (Hodge 1934, Hobbs 1943, 1947). A re-examination of the field area by Bretz et al. (1956) merely re-affirmed the hypothesis with the modification that there was evidence of repeated catastrophic flooding. Vindication came during a field meeting in 1965 when the participants asserted that 'we are now all catastrophists.' A recent re-assessment can be found in Baker (1973, 1978a).

The episode is notable for the fact that despite the outrageous nature of Bretz's hypothesis, and the attempts of his contemporaries to controvert it, it was eventually accepted, on the basis of the field evidence. It is note-worthy too that during the course of the protracted debate, a paper by Hobbs, the veteran glacial geologist, was rejected by two major journals on the basis of reviews by Bretz on the one hand, and Flint, of the 'opposition,' on the other. The rejections were based mainly on fact that the interpretations were inconsistent with the field evidence, and Hobbs eventually published his paper privately (Hobbs 1947).

A final piquancy can be added to this interlude by recalling that Sir James Hall (1815), in the paper reviewed in Chapter Six, inferred precisely Bretz's deep flood mechanism to explain streamlined forms west of Edinburgh. Baker (1973) showed how the streamlined loess hills matched lemniscate loops just as well as drumlins (Chorley 1959) so that although Sir James was wrong, in a very real sense he wasn't that far wrong!

THE CYCLE AFTER WORLD WAR II

The next chapter examines in detail post-War geomorphology, but cyclic notions persisted despite attacks from many quarters. It survived longest in the United Kingdom under the protection of Wooldridge. In Britain, cyclic thinking had metamorphozed into 'denudation chronology' in which a supposedly sinking global sea level was substituted for arbitrary uplift, partly under the persuasive influence of Baulig (1935). The pre-War classic by Wooldridge and Linton (1939) was revised and reprinted (1955, 1964) as was Baulig's 1935 monograph (1956). High late-Tertiary sea levels were related· to erosion surfaces at high levels, best preserved, and best known, on the Welsh massif (Brown 1960a). On the lower unglaciated south coast fragmentary benches, and scattered deposits, bespoke a sinking Pleistocene sea. Southern Britain was apparently ideally suited to Davisian analysis, but in the heavily glaciated terrain of the north the deposits were missing and the morphology was ambiguous.' When Wooldridge died, and the forceful imperative to find and study yet another upland mass was gone, regional Davisianism died thankfully. Yet as it did so, Brown (1960b, 1961) supplied a fascinating analysis that placed the evolution of Britain's scenery in a North Atlantic tectonic framework more explicitly than any previous work and revealed again the intimate links that regional landscape analysis must have with diastrophic history.

On the continent, post-War work in France swung towards the climatic approach, and although Birot (1960, Eng. trs. 1968) discussed these matters under the cyclic rubric, others did not. Cailleux and Tricart (1965) regretted the Davisian evangelism of Baulig, and Tricart, in addition to his climatic position, tried to revive a structural approach to geomorphology (1968, Eng. trs. 1974). The climatic school prospered in Germany under Budel and Louis, but, with the exception of a brief refinement by Spreitzer (1932), Walther Penck was neglected until the present day (Bremer 1983).

Lester King in South Africa maintained strongly cyclic concepts and his model for the parallel retreat of semi-arid slopes helped polarize the false debate between Davis and Penck because, on the one hand it was a particular realization of the Penckian model in which slope units did not become continuous, and on the other hand he claimed a world-wide application for the mechanics of his model, and the surfaces it generated. But both the diastrophic and the uniformitarian demands of such a global model were too extreme to be popular.

Isostasy again: the Cycle rejuvenated sui generis

Bowman (1926), concluding his review of Penck, thought that the future lay with the Davisian Cycle, modified in the light of isostasy. By the 1920s it was becoming clear that variable loading of the earth's crust had significant effects, and this was being illustrated very plainly by the accumulating evidence of postglacial isostatic rebound in Scandinavia, Scotland and Canada. In the longer run erosional unloading of the interior, and depositional loading of continental margins, either might trigger uplift and so generate new cycles, or greatly perpetuate the elapsed time to peneplanation (Lawson 1948). Nobody accepted the challenge until the 1950s when King (1956) and Pugh (1956) applied the idea to explain rejuvenation on portions of the African continent. Neglect continued and King himself scarcely gave the notion a passing mention in either edition of his monumental 'Morphology of the Earth' (1962, 1967). The obvious failing of cyclic schemes, of whatever hue, has always been that endogenous inputs are external to the model and arbitrary, indeed as Davis styled them, 'accidents.' Isostasy, at least in part, and especially in non-orogenic zones could supply rejuvenation sui generis, triggered from within the workings of the Cycle (Lawson 1948). Schumm (1963, 1976) has adopted the idea but it has not been used in any explicit fashion by regional workers. The idea came too late, and with insufficient precision, to save the Cycle, which was being savagely attacked on the mundane matters of diurnal processes, but it must be central to any eventual rebirth. At the same time it will have to be wedded to prevailing concepts concerning the vertical mobility of migrating plates (see Chapter 12).

A late twentieth-century perspective on the Cycle

The last word has not yet been said on the Cycle, in any of its many guises. For all its supposed faults, it has had a catalytic, rather than paralytic, effect on twentieth-century geomorphology. Opinions to the contrary forget that some similar alternative would likely have arisen to take its place if Davis had not developed it. Those who formally opposed it, and stoutly maintained the merits of their own schemes, have been subsumed within it by subsequent comment-ators. Climatic geomorphology has maintained some degree of independence by denying its diastrophic tenets, the control of base level, and the stablility of climate, but even so cyclic versions have been developed for most of the main climatic variants. Even the acyclic schemes hinted at by Fenneman (1936), and developed independently by Hack (1960, 1976), have had to move towards a partial reconciliation of their schemes with Davis'. It is still entrenched in textbooks, for want of a plausible alternative, but more significantly Carson and Kirkby (1972) deliberately adopted its schematic form:

> The peneplanation model may thus be challenged on a number of grounds, but, despite this, it seems that the Davisian system of successive upheaval and denudation of the landmass is probably the most appropriate framework for the study of slope forms. This is the system of reference used in this book.

The Gem of the Paha, Rockville, Delaware County.

Chapter Ten

THE IMPACT OF WORLD WAR II

> *Although geologic factors were important in military operations*
> *....the War was ended with relatively little awareness of this*
> *fact by the army.*
>
> *(C.A. Kaye, 1957)*

The Second World War was conducted on such a scale that it
had catastrophic impacts on science and society. Technology
applied to warfare came of age in 1939-45, and because a
large part of any non-nuclear war is played out on, or very
close to, the surface of the earth, an appreciation of
terrain in the light of the military machine quickly assumed
a paramount importance. The main theatres of European warfare
were already quite well mapped, due in part to the stimulus
of earlier wars, but the peripheral theatres of war were less
well-served: Asia, the Pacific islands, and North Africa. In
every case the cultural detail on existing maps needed
up-dating and almost everywhere there were lacking the
interpretative materials on the nature of near-surface
materials needed by military intelligence.

Most geomorphological work undertaken after the war
shows a marked divergence from the character of that
preceding it. Just as warfare makes technical demands, so too
it exposes people to the stimulus of new ideas and new
environments, and more than anything else, it breaks the
thread of continuity that inertia tends to promote in
academic thought and institutional life. It would be naive to
claim that all post-War innovations in the subject owe their
origins to wartime stimuli. In most cases a clear conceptual
inheritance from the pre-War period is easily detected; the
use of air photographs and the study of coastal processes are
just two examples. Nevertheless, the War did lead to radical
shifts in direction and marked changes of emphasis. A number
of emergent threads can be discerned, all of them intertwined
to some extent, although they are more distinct when separated by the Atlantic.

AIR PHOTOGRAPHS AND METHODS OF OBJECTIVE MAPPING

One response to military needs involved the use of air
photographs to supplement, or in poorly mapped areas also to

create, the bare topography of the map. The interpretation of air photographs, even those taken in ideal conditions, requires the experience of trained geographers and geologists, and one may surmise that many wartime air photographs were far from ideal. As W.W.Williams reported after the war (Williams 1947), groups of earth scientists were able to glean considerable information about enemy beaches by such methods, and air photographs, despite the many difficulties in obtaining them, were much used. Surprisingly though, not all the discussants of Williams' paper saw a great deal of academic future either in their future use, or in the strong flavour of the process-oriented work which had been done in wartime - studies had centred on wave dynamics and their relation to gradients, morphology and the particle sizes on beaches. Lt. Colonel Hart remarked that he couldn't see much practical application of the work. Vaughan Lewis on the other hand 'hoped scientists could still use air photographs in peace time and wanted to discourage the view that the photographic method is no good.' J.A.Steers, Chairman of the session, in his introduction, had remarked also that though the 'work was presumably not intended for other than military use ... clearly the information derived ... may prove of considerable value to research workers in general.'

The United States Armed Forces used geologists in essentially the same way as their British counterparts, for studies in terrain evaluation, water supply, military engineering and to assess beach-head landing sites. There were never very many geologists involved and they were usually attached to units that had a direct need of them (Kaye 1957, Simon 1957, Snyder 1957, Nace 1958). An account of the British experience can be found in King (1951).

The growing use of air photographs after the War was not especially controversial, although one inhibiting factor may have been the fact that, because many parts of Britain for long remained in military use, they were often unavailable or difficult to obtain from Air Force sources. Compared to published topographic maps at the same scale vertical air photographs carry orders of magnitude more detail and inevitably reveal the small scale complexity of landscape form and its intimate links to soil and vegetation. Such a revelation was perhaps irrelevant, perhaps unwelcome to the mainstream interest in denudation chronology in Britain after the War, even though the utility of air photographs had been specifically discussed as early as 1922 by Lee, and by Smith in 1941 on the basis of pre-War experience in North America.

The Linton school of morphologists at Sheffield

The relevance was not lost, however, on David Linton, Professor of Geography at Sheffield University and co-author with S.W.Wooldridge of the pre-War classic on the Structure, Surface and Drainage in South East England, (Wooldridge and Linton 1935, 1955). In a very enthusiastic paper, Linton (1946) laid out the basic principles of stereographic air photography and remarked that 'in a country of strong relief it affords a particular joy to the geomorphologist to fly around a polycyclic upland mass at "peneplane height" .. All critics of the concept of uplifted peneplanes should be flown around the western end of the Harz Mts. at 2000 feet.' The paper was based on wartime experience and the revelations of landscape it afforded was clearly responsible for the impetus and direction that Linton gave to the growing urge to find objective ways to represent landform.

At Sheffield, Linton and various colleagues, including Savigear, Waters, Straw, Johnston and Young, developed the idea and techniques of morphological mapping. The method is described in detail by Waters (1958) and is based on the field identification of slope facets and breaks of slope. In a similar vein, Savigear (1952, 1956, 1962) popularized slope profiling as a means of objectively recording and representing slopes. The method was developed independently by Fair (1947, 1948) and had been utilized even in the nineteenth century by Sorby (1850) and Tylor (1875). Nevertheless, it was first practised widely after 1950 as an aid to objectivity in the Penck/Davis controversy on modes of slope retreat. Young, originally a research student at Sheffield, subsequently refined the methodology of slope profiling in a series of papers culminating in a computer-based method called Best Units (Young 1971). Young's involvement with slope form led him to detailed work on the specific processes responsible for shaping the changing morphology of slopes. Consequently a fairly direct link between immediate post-War interests in objective morphology and the developing interest in surface processes in the 1960's can be established.

Air photography, terrain analysis and satellite imagery

In less well-mapped territory of enormous physical extent the practical use of air photographs was developed much earlier. In Canada the National Air Photography Library in Ottawa was founded in 1925 and by 1954 contained over two million prints. Parry (1967) in a review of Canadian geomorphology noted that this availability 'produced a revolution in geomorphological thinking in Canada, which marked a departure from the European tradition, with it's emphasis on detailed studies of small areas .. Air photographs provided a new perspective, ideally suited to regional physiography in general, and glacial geomorphology in particular.' These potentials were already realized by the mid-1950's: Chapman and Putnam's 'Physiography of Southern Ontario' (1951) involved field mapping at the 1" scale over 30,000 square

175

miles, much of it based on air photographs. In the period 1951 to 1966, Parry found an overwhelming interest in glaciation, periglaciation and regional physiography.

A related development in Australia was the development of Land System mapping. This was born partly of the necessity of mapping and assessing huge territories with low population densities and partly from the fears realized by the War in the Pacific. The basic notion was to categorize landscapes by means of topography and its related ecosystems. The first study, of the Katherine/Darwin region in North Australia, was begun in 1946 and the technique was developed to a fine art during the following decades and became increasingly dependent on vertical air photography. The contribution this made to geomorphology as a discipline came as much from the opportunities which it provided for staff and visiting geomorphologists to experience largely unknown landscapes (Driscoll 1964), as from the specific tool of mapping itself, though this should not be discounted as unimportant within the field of applied geomorphology.

Versions of Land System mapping subsequently diffused to other Commonwealth countries faced with comparable problems, mainly via the Overseas Development Ministry in the United Kingdom. A version specifically designed for military use, terrain analysis, took root in Britain and is reported in the work of Beckett and Webster (1965). It was tested experimentally in Uganda in 1969. Very similar survey techniques, aimed at fulfilling the same intentions, were devised by Soviet scientists (Isachenko 1965, trs. 1973, Chikishev 1970, trs. 1973), but these, and others in Eastern Europe (Klimaszewski 1956, 1961) have developed largely independently of experience in the West. The expansion of scientific research in the Canadian arctic described later in this chapter was also a variety of terrain evaluation; Bird's book on the physiography of the arctic (1967) was based on terrain descriptions sponsored by the RAND Corporation, and aerial reconnaisance combined with sample ground survey conducted from helicopter landings was an increasingly favoured technique (Bird 1959). Both terrain analysis and land systems mapping go beyond the mere morphological form and also concern themselves with the soils underfoot and their suitability to a variety of human uses. Once more a connection between morphology and an ultimate concern with process is established.

A conceptual predecessor to modern terrain analysis is possibly to be found in the work of Bagnold whose book on 'The physics of blown sand and desert dunes' (Bagnold 1941) was based on extensive military experience in North Africa and which has become a classic in process geomorphology. In the post-War period Bagnold has made very significant contributions to fluvial geomorphology (Bagnold 1960, 1966).

The War in the Pacific apparently had little immediate

effect on the use of air photographs in geomorphology in the United States, and this was also the case in Canada. Their use was already quite well established in the coterminous States (Lee 1922, Smith 1941) but the War did focus the attention of geologists on the morphology of coral reefs (Steers 1945, Fairbridge 1950a, 1950b) thus revitalizing an American interest in coral reefs dating back through Daly (1910, 1930), Davis (1928), Dana (1853, 1872) and ultimately to their first investigator, Charles Darwin (1842).

A comprehensive photo-interpretative review of Australian coral reefs by Fairbridge (1950b) was based very largely on wartime experience and airborne reconnaissance. A Great Barrier Reef Committee had been formed in 1922 and studies stretch back as far as Jukes (1847), but apart from this, detailed reef studies were lacking and many areas, for example North Australia, had not been visited by competent observers. This was rectified quite quickly when the U.S Military became interested in using Pacific islands for nuclear testing sites and a detailed study of Bikini atoll was made by Ladd et al. (1950).

Air photography has been, and will remain, a standard technique for earth scientists but the advent of earth orbiting satellites permitted a shift in spatial scale by several orders for aerial views of the earth. The potential of satellite photography was clear from the evidence of hand-held photography taken early in the 1960s, but the institution of systematic photography and the use of other wavelengths than that of visible light for remote sensing was delayed until the early 1970s. In 1972 the first of the ERTS (Earth Resources Technology Satellite, later termed LANDSAT) satellites was launched. Minor details are filtered out by the limiting resolution of the system so that the analysis of large scale spatial patterning becomes possible. It is not surprising, perhaps, that one of the first major studies using Landsat, supplemented by hand-held photography from the lower-orbit SKYLAB, was of the globe's major sand seas (McKee (ed.) 1979) since deserts are notoriously difficult for field workers and the mobility of the surface confounds the use of normal maps and surveys, even when they are available or possible. However, the technique has also proved useful in equatorial forests, for example the Amazon basin and augmented this time by side-scanning radar, where the forest environment and the vast scale militate against conventional operations (Baker 1978b). At the present time satellite remote sensing is the only spatial record available for land surfaces on the Moon and Mars, apart from maps made from the same images. On the latter in particular the analysis of remote imagery, tied to laboratory experimentation with plausible processes, is the only method of analysis available to geomorphologists unable to make field checks.

An inevitable by-product of the macro-scale analysis

forced upon users by LANDSAT imagery, a scale of 1:1,000,000 is used in McKee (ed. 1979), is a concern with regional patterning and presumably the long timescales which are likely to be involved in explaining the mosaics described. In short, regional physiography returns in a new guise.

COASTAL PROCESSES AND SHORELINE FORM

While the British environment may be a moderate one from the point of view of subaerial processes, it has an extremely active wave environment (Davies 1964), and there are many documented changes on its soft rock coastlines (e.g. Lyell 1830, de Boer 1964). Inevitably, therefore, an interest in the processes of erosion and deposition is to be expected. There was no shortage of nineteenth-century comment, yet by the early twentieth century, the overall emphasis was so regional that apart from Cornaglia (1889), for long unknown, Cornish (1898), and D.W.Johnson's book 'Shore Processes and Shoreline Development' (1919) there was astoundingly little published on the details of shore processes. Until the papers of Lewis (1931) in the United Kingdom and Evans (1942) in the United States there was but limited appreciation of the role played by waves in determining the configuration of the beach. Rather, tides and currents had been given pride of place even though the best nineteenth-century work had clearly identified waves as the primary agent of action on the shore (Hall 1843, Sandford Fleming 1854, Gilbert 1890).

The effect of the War

The impetus that the War gave to coastal research is difficult to assess, and might easily be over emphasized. During the thirties the work of Steers (1927) and Lewis (1931, 1932, 1938a) shows that active work on shore processes, and their relationship to constructional landforms, was well underway. By the beginning of the War wave tank research was being undertaken in an effort to resolve some outstanding problems (Ogilvie 1936, Bagnold 1940, Sweeting 1943), and Debenham (1942) records poignantly that the constraints of war meant the loss of the newly-built Physical Laboratory at Cambridge. A wave tank and a tide tank were two of the purpose-built units within it.
    Wartime needs apparently centred on the need to predict coastal behaviour at particular localities for military purposes, especially landings (Williams 1947, 1960), rather than on fundamental process research and this is understandable in the circumstances. In this sense the War caused a change of direction, away from a possible trend towards experimental work in the laboratory (Steers 1946, preface), and instead re-directed work towards a more

empirical approach to beaches and coastlines dictated by military needs, both offensive and defensive. Blackpool beach was used as a surrogate for the enemy-held Normandy beaches after the potential dynamics of the nearshore environment had been revealed by landing craft at Salerno, in Italy, which grounded offshore on bars they had cleared a few days earlier (Williams 1960).

Steers (1946) records that his book was already in typescript when he was approached by the Ministry of Town and Country Planning to make a survey of the entire coast, and his volume subsequently incorporated the additional information and experience gained in making the survey. An alternative view is that the experimental work may have turned out to be of limited value to coastal research and the Cambridge work might have developed just as strongly as it subsequently did along the lines of detailed, process-oriented field work. Certainly, it would appear from the post-War literature that laboratory experimental work is most easily adapted to the simulation of the dynamics of particular problems, and usually the scale and expense of these ensured that they have been carried out by engineers attached to well-funded institutions (Steers 1956).

Post-War coastal research

After the War, Lewis published virtually nothing on coastal geomorphology although a manuscript on Chesil Beach, long-awaited, was being developed at the time of his tragic death in 1961. However, C.A.M.King, a Cambridge graduate (1943) and a doctoral student with Williams (1949), revitalized experimental work (1949, 1951) and detailed process-oriented field work (1953, 1959), to become a leader in post-War geomorphology - a position that was acknowledged in a dedication to her of the Binghamton symposium volume on coastal geomorphology, (Coates ed. 1973). King was based at Nottingham University after 1952 and had diverse geomorphic interests. A significant portion of her coastal research involved developing a time series of beach profiles at East Coast localities, especially Gibraltar Point, (Barnes and King 1951, 1961, King 1970). In 1959 she published an influential text, 'Beaches and Coasts' that was quickly reprinted and which ran to a second and revised edition in 1972. After Bagnold's classic (1941) on sand movement, it was probably the first text in English to deal specifically with the dynamics of landforms in terms of processes measured in the field.

Complementary work with an emphasis on the movement of beach material was also undertaken by the Nature Conservancy and is reported in the work of Kidson (1950, 1959, 1960, 1963), Kidson and Carr (1959, 1962) and Carr (1965). This work also helped lead to an understanding of the morphogenesis of coastal landforms, in particular spits and bars, so that by the early sixties coastal form could be related to coastal process with some, if not complete, confidence (Hardy 1964, de Boer 1964). Tracer experiments performed with shingle (Kidson et al. 1956, 1958, Joliffe 1961), especially in the nearshore and offshore zone, had shown how insignificantly the movement of shingle contributed to the beach morphology. This work acted as a complement to beach studies of sediment movement in the inter-tidal zone.

Alongside the progressive work on coastal processes in the post-War period was an equivalent interest in the management problems of the coastal zone. Steers (1946, 1964) provided a major survey of the coast of England and Wales that was augmented, if not initiated, by the War. More popular accounts (Steers 1953a, 1966) aimed at a wider public, and the stimulus provided by the extensive damage accompanying the severe East Coast floods of 1953 (Steers 1953b) ensured that the need for proper coastal management was kept before the public, and the political eye during the fifties and early sixties when it might otherwise have been ignored. Official interest was also maintained by the work of the Physiographic branch of the Nature Conservancy. In the United States a Congressional mandate was given to the Army Corps of Engineers for shore protection, while on an academic front Louisiana State University established a Coastal Studies Unit under the directorship of R.J. Russell.

In the post-War period, cross-referencing between British and North American publications was more complete than it had been before the war, although the traffic was rather one-sided; British workers referring to American sources, but not vice versa. In North America, the Beach Erosion Board in Washington undertook experimental work in laboratory facilities (Keulegan 1948, Shepard 1950a, Rector 1954), until it closed in 1963, and perhaps fulfilled the role that the Cambridge facility merely might have duplicated, given enough time. In any case, King (1959) showed that there were severe limitations to the application of laboratory work to models of coastal processes. However, field observations on the United States west coast during the 1950s led to initial conceptual models of beach profile development in response to the seasons, and nearshore current trajectories more complicated than those employed in laboratory experiments (Shepard 1950b, Bruun 1954).

British work during the fifties also showed an awareness of continental research, particularly that of the French. At least two papers were written in French (Williams and King 1951, Williams 1953); there was a joint paper written by Guilcher and King (1962), and in 1958 an English translation and partial revision of Guilcher's respected 'Morphologie Littorale et Sous-marine' (Guilcher 1954) appeared, just before King's own book on beaches and coasts. Differences between the research of British, American and Continental workers of this period probably can be attributed more to the distinctive character of their individual coasts than to any fundamental difference in approach. Pre-eminently the fifties were a decade of basic field research; environmental stock taking and of experiential learning.

Hard rock coastlines

No mention has been made so far of research on hard rock coastlines. Steers, in the preface to his book (1946), remarks that paradoxically the most scenic coasts are those that hold the least interest physiographically, and indeed little work was done on such coasts. A paper by Arber (1949) emphasized the effect of geological structure and inherited interglacial morphology on the cliff profiles but the relatively short period for which sea level has been close to its present level means that in hard lithologies rather little erosion and cliff retreat has been accomplished. Curiously, the most significant work on cliff profiles was that of Savigear (1952) on a cliff in South Wales that was progressively removed from the effect of wave attack by development of a broad accreting beach. Savigear used this to demonstrate the decline in slope angle when the removal of talus from the foot of the cliff is inhibited, and his paper has become a classic in the field of slope studies. A later paper (Savigear 1962) attributed variable facets in cliff profiles to varying Quaternary sea-levels.

The other main feature of hard rock coasts is the development of the shore platform by wave erosion, but this attracted very limited attention (Hills 1949, Edwards 1951, Bradley 1958) until well into the sixties. The Pleistocene corollary of shore platforms is raised beaches and these did interest some workers, but mainly from the point of view of Quaternary chronology. The really significant work in this field dates from the remarkable series of studies begun by Sissons (1962) and still in progress (Sissons 1982).

Tropical coastlines

The emphasis so far has been towards mid to high-latitude, and primarily northern hemisphere, beaches. Some mention was made in the last section of the renewed impetus given to studies of coral reefs by the War and this was then maintained by the use of Pacific islands for nuclear testing. However, tropical coasts tended to lack the background of nineteenth-century studies typical of many North Atlantic coasts so that a considerable degree of stock-taking was necessary before significant advances could be made. To a large extent the general lack of work on tropical coasts, until the sixties, reflected the status of most tropical

countries as generally under-developed, with a consequent
lack of qualified personnel to carry out research, whether in
research institutions or in Universities. The situation is
exemplified by Gregory (1962) who reported raised beaches
from Sierra Leone which he could only tentatively correlate
with the Mediterranean, and by Carter (1959) who, in a study
of mangrove succession and coastal change in south-west
Malaya could only reference Johnson (1919) apart from a
variety of authorities local to the area.

THE NEW FLUVIALISM

The previous two sections have dealt with themes that were
intensified or deflected by the events and experiences of
wartime. In contrast, fluvial geomorphology, taken as the
broad category of landscape processes due to water flow over
a free surface, was initially unchanged by the events of war.
Yet, within a decade, the entire character of the field had
changed from that of a subject concerned with an essentially
regional approach to one increasingly dominated by experimental activity, both in the field and in the laboratory.
During the fifties this paradigmatical shift was largely
confined to the United States, although the independent work
of Sundborg (1956) in Sweden is a significant exception, but,
with the publication in 1964 of 'Fluvial Processes in
Geomorphology' (Leopold, Wolman and Miller), the new
methodology reached a world-wide audience, at least of
English-speaking geomorphologists, as work in the United
States slowly matured after more than a decade of innovative
activity.

Horton and Strahler

It is never easy to say what influences or papers have
triggered new ideas, but it is generally agreed that one of
the most influential papers in the immediate post-War period
was that by Horton (1945). It was entitled 'Erosional
development of streams and their drainage basins;
hydrophysical approach to quantitative morphology' and was
prefaced by the remark that 'in spite of the general
renaissance of science in the present century, physiography
still remains largely qualitative.' It mentioned also that
the important topic of channel development would not be
considered in detail: the paper mainly considered the effect
of overland flow on erosional development and the statistical
structure of drainage basins viewed as areas and networks.
The importance of Horton's paper was recognized very quickly:
Bryan, immediately after the War, insisted that his students
take Horton seriously (Stearns, personal communication 1982)
and by the time of Bryan's death in 1950 the first papers of

Strahler were beginning to appear advocating dynamic geomorphology, an approach to landforms based on general physics (Strahler 1950a, 1950b, 1952). This advocacy was mirrored by a similar French interest that established the journal 'Revue de Geomorphologie Dynamique' in 1950 under direction of De Margerie, De Martonne, Cholley, Cailleux and Tricart, a mixed bag of Davisian and climatic geomorphologists. The potential importance of Horton's paper already was noted in print by Carlson (1950), but this sudden recognition merely emphasized the neglect of his, and similar, work in the pre-War period (p162). Horton's 1945 paper listed nine previous papers by him that cover essentially the same material, all readily available in the civil engineering and soil science literature. Once more there recurs the theme of civil engineers pioneering advances in fluvial understanding, echoing the neglected notebooks of da Vinci, the eighteenth century French and Italian canal engineers and inevitably, Gilbert, at heart and in method an engineer, from the turn of the last century.

Two main lines of work have their spiritual heart in Horton's paper. One is, as mentioned, the series of papers emanating from Strahler and his disciples based at Columbia University. These were, in many ways, complementary to the interest in England in finding objective methods of representing surface form. Part of the initial Strahler thrust was directed at testing quantitatively fundamental Davisian assertions about slope angles, but in time the group developed very complex methods of drainage basin representation and classification and, in the hands of Melton (1958a, 1958b, 1960), went about as far as was possible in the pre-computer era with the categorization of basin types in relation to measures of geology, vegetation and climate. Most of the analysis was based on quantitative data measured from maps and aerial photographs although in some instances field work was used to collect or check data at selected sampling points. The method became known in the English language as morphometric analysis although the term was already in use in French and German, and the methodology had been developed extensively by the French before the War, Baulig being particularly active in applying the technique to his studies of regional geomorphology (Baulig 1926, 1939, 1959, de Martonne 1940, Peguy 1948, de Smet 1951). There was, however, a change of scale. In general, earlier applications were at the national or regional scale, and directed towards terrain and river profiles, whereas the Strahler school worked with the largest map scale available, preferably in the region 1:25,000 and with a greater variety of variables. Clarke (1966) dubbed the Strahler school as micro-morphometry but the term is not in current use.

Apart from their theoretical interest, morphometric methods have provided useful inputs to quantitative models of flood prediction in basins bereft of data (Wong 1963). This aspect of their use has continued to generate research in the field of network morphometry, although their value to geomorphological, as distinct from hydrological studies must be placed in perspective. Stoddart (1969) remarked on the fact that neither the French nor the German climatic geomorphologists had used the standard techniques to develop or substantiate their arguments. Undoubtedly this reluctance stems from a distrust both of the basic data sources and the lack of inferential power which bedevils many morphometric measures. Topographic maps may be particularly unreliable in tropical areas, and especially in the presence of a heavy vegetation cover (Eyles 1966). It has not proved easy to relate measurements to environmental controls, despite careful statistical controls and even their proponents have remained cautious (Melton 1958). In the more general sense of quantitative measures taken from maps, or occasionally photographs, morphometry remains a valuable adjunct to geomorphological enquiry. However, like quantitative methods, they are seen to be most useful in conjunction with other techniques, in a supportive role, rather than in isolation where their usefulness is more readily queried.

The other thread of activity was that taking place in the United States Geological Survey. Flood Control programs in the United States began in the thirties with the establishment of the Tenessee Valley Authority (1933), but the combined effect of the Depression years and the Second World War was to delay the real momentum of such programs until after the War. Stream gauging is expensive when a dense network of stations is envisaged, and at least some of the geomorphological work that came out of the Survey was done with the intention of trying reduce the complexity of natural channel hydraulics to some sort of predictable order. If this could be done, then a relatively sparse network of gauges would suffice to provide the required hydrological information. Horton's paper had, in a sense, made order of the converse of this problem: the physical behaviour of the natural unchannelled surface; or had indicated how this behaviour might be quantified. The work of the Leopold school, particularly as summarized in their 1964 text, was much stronger on the channel hydraulics than on the behaviour of the land surface. Indeed they did not cite any Professional Papers on this subject and relied instead on a variety of pre-War and post-War work of which the most notable are perhaps Gilbert (1909), Bryan (1923), Horton (1945), Schumm (1956) and Young (1960). This is not to say that they did no work in this area; they did, but it is in the nature of the problem that useful results are more slowly achieved than in the field of channel behaviour, if only because the intensity with which the processes operate is that much weaker since they are being diffused over an area rather than concentrated in a channel. Significantly, the best work of the Survey in this field lay in the semi-arid areas of the American South-West (Schumm and Hadley 1957, Leopold et al. 1966). A much later text by Carson and Kirkby (1972) had the specific intent of filling the gap left by 'Fluvial Processes' in the matter of natural slopes.

The Leopold group centred on the U.S. Geological Survey

The fundamental paper emanating from workers in the United States Geological Survey (USGS) and their associates was that of Leopold and Maddock (1953) on 'The hydraulic geometry of stream channels and some physiographic implications.' The title introduced the new term 'hydraulic geometry' to geomorphologists but also indicated that the authors were concerned with quantifying landscape within the existing rubric of geomorphological theory, at least as far as this was possible. Although they could not address regional and cyclic problems directly they remained aware that this was the ultimate objective (Leopold et al. 1964).

Hydraulic geometry interrelates measurements of the channel with its water and sediment discharge using the mathematical apparatus of power functions. A paper by Leopold alone also appeared in 1953, on downstream changes in the velocity of streams; an attempt to correct the prevalent and qualitative notion that the velocity of streams decreased systematically in the downstream direction because of decreasing slope. This idea, which is quantitatively imprecise as stated since it is not clear what measure of stream velocity is being considered, often is attributed to Davis and was certainly assumed to be part of the Davisian explanation for floodplains and meanders (Russell 1898, Milne 1928, p92), although Leopold's paper does not specifically attribute the statement to Davis or any other pre-War geomorphologist. It is but one thread of interest in the new view of river morphology but it serves to illustrate the methodology, and its uneasy relationship to classical geomorphology. Leopold showed that on average, over a collection .of rivers, mean velocity measured at a discharge corresponding to the mean annual discharge, increased rather than decreased in the downstream direction, and that a variety of other adjustments could be made by a channel to compensate for downstream decline in stream gradient. The downstream increase in velocity is often only a slight one and subsequent re-evaluation suggests that perhaps it is better regarded as constant in the downstream direction (Carlson 1969). The finding came as a first rate shock to most traditional geomorphologists (Mackin 1963) and dramatized the fact that few of them had ever measured the mechanisms of geomorphological processes in the field; a point that was perhaps more pertinent than the nature of velocity change itself. Recall, too, that Leopold was merely rediscovering a point that Lyell (1830) had borrowed from French engineers (p94).

Ironically, it may well be that Davis, who was extremely well-read in all his contemporary geological literature and especially that of the USGS geologists, got the idea from Gilbert's classic study of the Henry Mountains (Gilbert 1877). The following quotation occurs in Chapter V of the report, at the end of a section entitled 'Transportation and Comminution.' Gilbert gives a thorough discussion of the relation between the amount of sediment that is carried by a stream and the calibre of the sediment and concludes that:

> the velocity of a fully loaded stream depends (ceteris paribus) on the comminution of the material of the load. When a stream has its maximum load of fine detritus, its velocity will be less than when carrying its maximum load of coarse detritus; and the greater load corresponds to the less velocity (Gilbert 1877, p107).

Gilbert does not specifically say that velocity decreases downstream; but for many streams, and in the statistical spirit of the Leopold school, that would be a perfectly fair inference to make since, generally, sediment size decreases in the downstream direction.

This discussion of velocity illustrates the general character of thinking in the Leopold school, and this is now more important than the actual findings themselves which have long been refined, generalized or superseded. A retrospective view of this new thrust was given by Leopold and Langbein (1963) at about the time that their 1964 text was going to the printers.

> At any time the need for a set of questions, implicit or explicit, is paramount. Over and above that, there is a time for raw data and there is a time for new theory. Progress depends on both. For several decades governmental authorities had been collecting data on rivers. No one knew just how to apply this store of information to geomorphic inquiry. No one knew what questions to ask. Then in 1945, Horton set forth a new hydrophysical theory of landscape in contrast with the anthromorphic word pictures of William Morris Davis. The analysis of river data began soon after. There followed a decade and a half of analysis using the data available in conjunction with ideas stimulated by Horton's theory. Not much more is likely to be gleaned from either. The time is set for new theory and new data.

From this it is clear that one reason that comparable studies of slopes lagged behind was that there was no comparable bank of published data. They lacked also the public focus that attends the flooding of densely populated areas; and to the public eye, catastrophic slope failures appear to be much less susceptible to solution by programs of public policy such as flood plain zoning and land-use control.

Leopold and Maddock (1953) established the methodology of the new fluvial geomorphology but this work and that descending from it over the next ten or fifteen years, failed to solve any of the fundamental problems of regional and cyclic geomorphology. This was essentially because the time scales over which the two fields of study operate are so vastly different: the one sensitive to daily changes in flow and with documented histories scarcely reaching beyond a few centuries, the other concerned with periods of thousands to millions of years. The problem caused by such disparities was not clearly enunciated until the 1965 paper of Schumm and Lichty on 'Space, time and causality in Geomorphology,' which will be reviewed in the next chapter.

Medium-scale fluvial problems

Progress was made on some medium-scale problems. River meanders attracted considerable theoretical and practical work with papers by Bagnold (1960), Leopold and Wolman (1960), Langbein and Leopold (1966) and in the context of valley meanders by Dury (1954, 1964a, 1964b, 1965). Dury's interest in the sweeping geometry of valley meanders began in the Cotswold Hills in England and the work subsequently benefited from his period of assignment to the USGS in 1961/2. Valley meanders are meanders on a scale one or two orders of magnitude larger than the river channels they contain and appear to originate from the peculiar hydrological conditions of the late Pleistocene and the early Postglacial periods. Without doubt an understanding of their genesis would forge a valuable link connecting the hydraulic geometry of alluvial river channels with the excavation of bedrock channels and valleys.

Another medium-scale problem was that of floodplains, and it proved possible to connect the periodicity of flooding with the processes of floodplain construction and the movement of river meanders in a reasonably satisfactory way, (Wolman and Leopold 1957, Leopold and Wolman 1957). Historical perspective was added to this from studies of river trenching and aggradation following the construction of dams, and from nineteenth-century environmental disturbances in the American west which were dated by chronologies based on Indian archaeology. However, compared to the spate of work on river channel dynamics the historical material, almost anecdotal in character, was insufficient for the development of theory and this is reflected in the relatively short space devoted to it in their 1964 text.

Theoretical developments

The quotation given above from Leopold and Langbein mentions the development of theory in addition to the collection and analysis of data. Several of the specific papers already quoted have a considerable theoretical content, but the more general principles underlying the papers on particular landforms are best expressed in the paper by Leopold and Langbein (1962) on the concept of entropy in landscape evolution, in which essentially statistical models are proposed, based loosely on thermodynamic models from general physics. Later papers by Chorley (1962, 1965) placed these thoughts within the then popular framework of General Systems Theory and specifically launched an attack on Davisian models of landscape evolution by equating their philosophy to modes of closed-system thinking, and to which one British economic geographer felt moved to respond, so deeply rooted was the Davisian model in standard geographical thinking (Chisholm 1967). Interestingly, Culling, working independently in Britain, was able to use essentially identical formalism with rigorous mathematics to articulate the Davisian scheme (Culling 1957, 1960, 1963, 1965). Regrettably, his work was largely ignored at the time that it was published and it had very little apparent impact on contemporary debates.

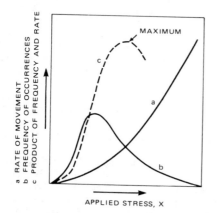

One theoretical paper of more immediate practical import was that on the 'Magnitude and frequency of forces in geomorphic process' by Wolman and Miller (1960). They pointed out in this paper that the bulk of geomorphic work is carried out by fairly frequent events of medium magnitude which, over a period of time, far outweigh the impact of occasional very high magnitude events on the one hand, and the negligible effects of extremely high frequency, and yet very low magnitude events on the other hand (Figure 15).

Figure 15. *Relations between rate of transport, applied stress, and frequency of stress application. Redrawn from Wolman and Miller, 1960.*

It is probably fair to say that over the last two decades, that paper has had an enormous effect on general thinking. The principles are now used in routine analysis and are classically exemplified in the landscape by the identification of the bankfull discharge as the most formative and effective flow in the hydraulic geometry of channels and channel patterns. The fact that this particular discharge has been so frequently assailed in the literature, and other types of mean flow suggested in its place, e.g. the mean annual flood, merely indicates the entrenched position that it has now achieved.

While the new fluvial school could not solve the classical problems of cyclic geomorphology it did suggest an alternative. Hack (1960) adopted dynamic equilibrium as the central notion of the new fluvial school (although it's philosophical antecedents go back to Gilbert's study of the Henry Mountains, and to Fenneman (1936)), and used it as the basis of a general theory of landscape development in which the landforms found are essentially independent of time. The idea was specifically applied to the Appalachian landscapes and implied that the landscape maintains its form during downwasting and never develops the final stage of the Davisian cycle, the peneplain. King (1953) had termed the peneplain 'an imaginary landform' and Hack observed that typically Davisian maturely dissected terrains were found in abundance and might therefore be taken as the landscape norm, with spatial variations in surface form and slope being responses to differing lithologies and their variable modes of weathering, a notion very similar to Fenneman's description of non-cyclic erosion (1936), although Fenneman is not referenced by Hack.

Hack's paper had a mixed reception and Mackin thought, semi-seriously, that the paper was written merely to annoy him (personal communication 1968). It was important, however, as the first acyclic scheme to be taken seriously, and at least indicated the character such a model would have to have. Its difficulties, as Hack admitted, were that whereas within the Davisian scheme the idea of dynamic equilibrium fitted uncomfortably, within the Hack model the persistence of relic topography, not smoothly graded to the rest of the landscape, implied that 'changes in topographic form take place as equilibrium conditions change' so that if, for example, 'the relative rates of erosion and uplift change .. then the state of the balance or equilibrium constant must change. The topography then undergoes an evolution from one form to another' and 'if, however, sudden diastrophic movements occur, relict landforms may be preserved in the topography until a new steady state is achieved.' To the cynic this sounds suspiciously like a change between partial cycles in the Davis model, themselves induced by exogenous energy inputs. Hack argued, however, that in the Davis model exogenous inputs caused parts of the landscape to be out of equilibrium for considerable periods of time whereas under the dynamic equilibrium model all parts of the landscape are simultaneously in equilibrium although they may be adjusting their form to the new conditions.

Hack's radical reassessment was based not merely on a new view of exogenous processes but on the knowledge that 'cyclic theories of landscape origin are close relatives of the theory of periodic diastrophism.' Hack rejected cyclicity and adopted Gilluly's equally radical view (1949) that:

> diastrophism has not been periodic but was almost continuous through time, though the form and location of diastrophic processes has continually changed. This concept of continuity of diastrophic processes is, of course, discordant with cyclic geomorphic theories, but is in harmony with the equilibrium concept outlined here.

It is perhaps not surprising that recent assessments of competing models of landscape evolution have seen ways of incorporating dynamic equilibrium within the notion of multicyclic schemes (Palmquist 1975, Higgins 1975) and even newer paradigms of diastrophism, although one difficulty with any model of landscape evolution, or lack of evolution, is ensuring that it is falsifiable. Bishop (1980) has attributed the longevity of the Davis cycle to its intrinsic unfalsifiability, and the same criticisms probably apply to the Hack model even if it has been less widely adopted.

GLACIAL AND PERIGLACIAL RESEARCH

The fields of glacial and periglacial geomorphology show a mixed response to the War. Glacial research shows virtually no impact perhaps because the surface of continental ice sheets was of limited territorial and strategic significance quite apart from their forbidding environment. In contrast, post-war interest in periglacial, and this meant essentially arctic and sub-arctic terrains, was certainly promoted by the geotechnical problems encountered in large scale engineering in such regions, itself prompted by strategic considerations.

Glacial geomorphology

The cyclic idea which so bemused fluvial geomorphology remained essentially a pedagogic device when applied to glacial landforms, and had nothing to say about the depositional landforms of the great Pleistocene ice sheets. After the recognition of multiple glaciation in North America following the controversy between Chamberlin and Wright (Schultz 1983), and in the Alps (Penck and Bruckner 1909), both of which confirmed James Geikie's interpretation of the evidence, progress in glacial geomorphology was steady, if

190

unspectacular. The school of glacial protectionists (Garwood 1910) never affected majority thought although it is now recognized that glacial erosion, however real, may be highly selective (Linton 1963). Hobbs (1911) drew attention to the importance using polar, as well as alpine analogues for understanding Pleistocene ice sheets, and he stressed particularly the difference in behaviour between ice sheets and valley glaciers. The period from 1890 to 1920 was a particularly rich one for glacial studies with an especial focus on Alaska (Tarr and Martin 1914, Muir 1915). However Polar expeditions at this time provided little information that was of value to geomorphology.

The mapping of glacial landforms, especially those resulting from deposition, is slow and the delineation of the marginal positions of the ice sheets depends on an accurate assessment of their meaning, in terms of the marginal processes of glacial retreat. Despite the need for precise interpretations most of this work was carried out in formerly glaciated regions, rather than in areas with active ice. The majority of papers on glacial topics between 1900 and the 1960s, when they are not regional, concern the classification and interpretation of specific landforms, such as eskers, kames, drumlins, drift stratigraphy, and meltwater channels. A fundamental idea in this work was provided by Flint's distinction (1929) between active and dead ice and their characteristic assemblages of landforms developed during glacial retreat. He applied the idea to glacial retreat in Ireland (Flint 1930). Similar views were reported by Andersen (1931) from Denmark, but the notion did not go unchallenged (Lougee 1940, Flint and Demorest 1942). In terms of scale, work was primarily on drift stratigraphy and regional patterns; e.g. drift boundaries (Flint 1943), the delimitation of the isostatically deformed shorelines of the Great Lakes, and their relation to the retreat of the Laurentide Ice Sheet (Hough 1958). A technical paper of considerable impact was Holmes' definitive use of till fabric to elucidate the varying motion of the now-vanished ice sheet (Holmes 1941, Dreimanis et al. 1957).

The lack of impact of the War on general thinking can be judged from the references cited in Thornbury (1954), the most significant post-War geomorphology textbook before the 1960s. In the three chapters devoted to glacial topics a total of 82 references are cited with dates before 1945, only two of which predate 1900. Post-War references number 39, admittedly only an eight year period. Flint (1957) shows a similar structuring. However, when a subject is undergoing rapid change, citations tend to be strongly biased towards the recent period.

The smouldering controversy during most of this period was the problem of the Channeled Scabland in eastern Washington, reviewed in the previous chapter. Thornbury

(1954) gives an uneasy, though unbiased, review of the subject. He endorsed no particular theory, but he clearly harboured catastrophic doubts. Flint (1957) ignored it, while additional field investigation by Bretz et al. (1956) merely re-affirmed it. These attitudes serve to confirm that continental glaciers were still viewed primarily through the stratigraphy of their deposits, rather than through the astounding energetics that their former presence and known dynamism implied.

Notable books on glacial geology have appeared at roughly ten year intervals following W.B.Wright's 'The Quaternary Ice Age' (1914, 1937), R.F.Flint's 'Glacial Geology and the Pleistocene Epoch' (1947), and 'Glacial and Pleistocene Geology' (1957). The same year saw the publication of J.K.Charlesworth's encyclopaedic two volume 'The Quaternary Era.' None of these books, however, signal radical shifts in thinking about glacial geology and landforms, although there were significant advances in the areas covered and in the understanding of specific features.

In Britain, for example, the post-War period was marked by the understanding that glacial erosion and sub-glacial meltwater could act in concert to cause drainage diversions previously thought to be river captures (Linton 1949, 1951, Dury 1953). Peel (1949), from careful surveys, showed that some of the meltwater channels in Yorkshire with 'humped' long profiles resulted from sub-glacial streams, and not from the sequential overflow of ponded glacial lakes as first described by Kendall (1902) in the Cleveland Hills. Vaughan Lewis, the Cambridge geomorphologist who turned to glacial topics from earlier research on coasts, was responsible for inspiring work on glacial processes amongst his students, long before it was generally popular. His field work took him to Iceland (1938b, 1954) and Norway (1960 ed.), and he was a founder member of the British Glaciological Society, whose 'Journal of Glaciology' carried significant papers on glacial physics and especially the mechanics of ice flow (Nye 1959, 1965, Glen 1952, 1958) and upon which was founded the revolution in glacial geomorphology of the late 1960s.

Scandinavia in the post-War period provided some significant new ideas to the debate about the mode of retreat of ice sheets. In mountain areas extensive till or fluvioglacial deposits are often missing and Mannerfelt (1945, 1949) was able to show that meltwater flowing close to the ice surface had eroded bedrock channels across divides, or along the contour, to provide a record of ice sheet thinning. Hoppe (1957) also supported these approaches and Hollingworth (1952) and Sissons (1958) introduced the methodology to mountain Britain, as a valuable adjunct to existing techniques and a radical alternative to the entrenched Kendall model of overflowing lakes, already attacked by Peel (1949).

Exploration of new terrains has always proved a stimulus to the understanding of landforms and the most significant feature of the post-War period for glacial studies may have been the establisment of permanent scientific bases in Greenland and Antarctica from which a knowledge of large ice sheets could be gained. The cumulative effect of this work was bound to be slow but by the late fifties and early sixties the results were beginning to affect both glaciological theory and thinking on glacial landforms (Schytt 1956, Bishop 1957, Hollin 1962, Zotikov 1963, Hansen and Langway 1966). A complementary exploration was also being conducted over the formerly glaciated terrains of the Canadian north (see the next section) so that the nature and behaviour of continental ice sheets began to be comprehended both from above and below.

The final acceptance of Bretz's Spokane Flood for the Channeled Scabland came during a field meeting of an international conference on the Quaternary (Richmond 1965). At last a 'fast' catastrophe could be accepted within the rubric of a 'slow' catastrophe and Malde (1968) later traced similar 'large' flows to overspills from Lake Bonneville.

Periglacial geomorphology

Lozinski first introduced the term 'periglacial' in 1909 to designate terrain that is peripheral to glaciers, either spatially or climatologically. The recognition that certain rubbly slope deposits might have originated in a cold climate certainly goes back to James Geikie (1874) while the widespread existence of permanently frozen ground was first established during the Russian settlement of Siberia in the second half of the nineteenth century. An Institute devoted to the problems posed by such terrain was established at Yakutsk in Siberia well before the War. Serious work on periglacial landscapes in North America was first undertaken as a result of the engineering problems encountered in the construction of the Alaska Highway during World War II and the subsequent interest in the arctic as the first line of defence in the Cold War. Research work in the arctic was stimulated and funded by the Arctic Institute of North America (comprising Canada, Greenland and United States), the Defence Board of Canada, and RAND Corporation (on behalf of the U.S. Air Force). According to Parry (1967) published work on the arctic doubled after the War and that there were twenty one funded projects in the period 1945-55, and thirty one in the subsequent decade. By the late 1950s over half the published work on geomorphology in Canada concerned the north.

The vast scale and general inaccessibility of the North American arctic lands meant that a useful perspective on the work achieved required at least a decade of accumulated inquiry, although the main features to be expected had been summarized by Peltier (1950). The first integrated accounts began to appear towards the end of the fifties (Bird 1959, 1967, Hamelin 1961, Washburn 1956) and the first detailed map of permafrost in Canada was published by Brown (1960, 1967, 1970). The term 'permafrost' was introduced by Muller (1945) to indicate permanently frozen ground (a definition by temperature not necessarily implying the presence of ice), but its exact delimitation and relationship to past and present climate is a matter requiring much patient fieldwork and borehole information. Its presence at depth is not revealed directly by air photography, and paradoxically its former existence may be easier to detect, from the evidence of patterned ground, than its present extent.

The post-War period was therefore characterized by the necessities of regional physiography in the north; discovering what was where. The hugeness of the scale is indicated by the statement that during one summer in the fifties trimetrogon aerial photography covered 950,000 square miles (Parry 1967). Eventually a very significant outcome of this routine and necessary exploratory work was Andrews' elucidation (1970) of the main features of post-glacial uplift in arctic Canada from a consideration of raised shorelines, glacial limits and radiocarbon dating. The date of this monograph indicates the scale of the logistic problems that had to be overcome in order to write it.

In Europe careful field work on periglacial slope processes was instigated by Rapp (1959, 1960a,b, 1962), Rudberg (1962) and Williams (1957, 1959) in investigations of the types and rates of mass wasting found on mountain slopes in Lappland, Spitzbergen and Norway. Elsewhere in Europe, and in the temperate latitudes of North America, the former presence of permafrost and solifluction was deduced from patterned ground revealed in air photographs (Williams 1964) or from relict slope deposits. In Britain a lively controversy raged around the possible periglacial origin of tors (Linton 1955, Palmer 1956, Palmer and Radley 1961, Linton 1964). An academic forum for the burgeoning periglacial interests eventually was provided by the International Permafrost Conference which has met every ten years since 1963. However, the field had had its own journal since 1954 when the journal 'Biuletyn Periglacjalny' was started in Poland.

KARST AND THE LANDFORMS OF LOW LATITUDES

Karst geomorphology

The cyclic notion, as indicated in the last chapter, was adapted to karst terrain by several people independently.

Because it was merely a lithological variant of the fluvial cycle this was to be expected. However, until 1930 the actual mechanics of cavern solution, the crucial factor in developing karst terrain, were only vaguely understood; it was assumed to be the inevitable consequence of groundwater circulation in soluble bedrock, and in cyclic terms it was a one cycle theory. Davis (1930) proposed as an alternative that cavern development was a two cycle process, with cavern solution by phraetic waters in the first cycle, and precipitates forming in the caverns during the second cycle. Bretz (1942, 1953) was a vigorous supporter of this view, and Sweeting (1950) showed the coincidence of cavern levels and erosion cycles in the limestones of Northern England.

Limestone terrains are distinctive because of their susceptibility to solution and the fact that solution is the overwhelmingly dominant process in their formation. Corbel (1957, 1959), in the light of these facts, and the knowledge that solutional loss was an under-rated component in other regions, published a provocative paper in which he claimed that solutional erosion ought to be enhanced in cold humid climates and at high altitudes because of the greater solubility of carbon dioxide in cold water. Sweeting (1964) pointed out that this ignored important secondary factors such as the much greater availability of humic acids within the soils of temperate regions, and the total quantities of water which might be available in unit time. Exactly the opposite view was published by Lehmann (1964), one of the leaders in the German school of climatic geomorphologists, who believed that it was the humid tropics which underwent the greatest solutional erosion.

These two contradictory viewpoints, undoubtedly based on an over-simplified understanding of the many complications surrounding the chemical mechanism of solution in the landscape, led to many subsequent studies of solutional rates. The basis of these studies was the hydrologic system, often underground, that evacuated limestone regions, and the developing interest in karst hydrology during the fifties was a helpful precedent for later process studies in fluvial geomorphology. A recent assessment of the contradiction posed above, and it was not an issue that could be resolved quickly to general satisfaction, suggests that neither statement is correct (Smith and Atkinson 1976). Nevertheless, the process of resolving the problem has led to a much more detailed knowledge of global solutional processes and their relation to mechanical erosion.

An equal amount of uncertainty encircled the debate about whether karstic landscapes are significantly and necessarily distinctive in different climates. The earliest detailed work from a strictly climatic angle dates from Lehmann's study of Java (1936) and his work, and that of his pupils led him to claim (Lehmann 1954) that karstic landforms delineated 'clima-specific' regions. However, as with any system of classifications, there were bound to be problems of definition and description. Smith and Atkinson note that little use was made of Hortonian morphometry to describe limestone terrains until Williams (1966, 1972) outlined and illustrated suitable modifications. They also remark that although mean annual runoff is now usually recognized as the main control on erosion rates, with temperature as an important secondary factor, it is still the case that:

> the study of process has not been of great help in explaining the reasons why some distinctive landforms and landscapes occur only in the Tropics, whereas others are found throughout the Temperate and Tropical zones. Part of the difficulty lies in defining the problem...With only verbal tools of analysis at our disposal, it has proved almost impossible to describe and distinguish different sub-types of karst objectively.

They hope therefore that some resolution of the problem can be attempted through the use of morphometry as the basis of terrain description, a prescription more radical than any yet proposed for 'normal' landscapes. However, the same point was made in discussion by Stoddart (1965) ten years previously so that progress in the matter obviously has been slow.

The landforms of low latitudes

Low latitudes provide a radically different physical environment to that of the temperate latitudes from which most geomorphologists originate. In the two decades after the War many countries in low latitudes were under a colonial rule from which they emerged to Independence during the sixties. Research work emanating from these regions usually came from the newly-founded universities, or from the officers of colonial (later national) geological or land resource surveys. There is on the one hand a reasonable abundance of literature, but on the other hand the literature itself covers an infinitesimal fraction of the available landmass. The problem therefore is one of representative coverage. The climatic school of geomorphologists, primarily the Germans and French, resolved the problem by maintaining the regional scale of enquiry, and in a manner akin to the national terrain surveys, they tried to identify widespread systems of environmental processes which would yield distinctive landscapes (Budel 1948, 1957, 1963, Tricart and Cailleux 1965). Cyclic geomorphologists took a similar approach (King 1953, 1963, Birot 1960, Eng. trs. 1968), but stressed the controls of base level and diastrophism. Specific work on processes was very limited in low latitudes but there was an emphasis on the pedogenesis of the deeply weathered profiles overlying the basal surface of unmodified rock (Ruxton and Berry 1959).

The significance of changing climates, especially during the Quaternary, to understanding regional landscapes was gradually recognized in the post-War period. In a general way it was realized that 'pluvial' periods in the tropics probably corresponded to 'glacial' periods in high latitudes, and the classic work was that done in East Africa on the levels of the Rift Valley lakes (Nilsson 1931, Wayland 1931, 1934, Leakey 1936) although it is now known that this record is disturbed by Quaternary tectonics and volcanism. Detecting the effect of climatic oscillations on humid landscapes is very difficult, due in part to the problematical inter-relations of water and sediment discharge on slopes and in channels, but one simplification can be achieved by looking at areas which are presently extremely arid, and devoid of fluvial action. For this reason the increasing attention paid to tropical deserts during the 1950s and 1960s is of especial importance (Hack 1941, Price 1950, Peel 1966).

The southern margins of the Sahara proved to be a particularly fruitful locality, and Grove was able to show (1957, 1958, 1959, Grove and Pullan 1963, Grove and Warren 1968) not only that Lake Chad had once been twenty times larger, but also that active desert dunes, now fixed by vegetation, had previously extended south far beyond the present limits of the desert. As aerial, and eventually satellite, photography became available it proved possible to map the regional patterning of active and fossil dune lines and either to relate them to the sparse meteorological records, or to infer paleo-wind directions. Thus it could be shown that arid terrains had been both more and less extensive in the past, and Butzer (1958) was able to deduce the position of isohyets from the fauna shown in the cave drawings of the Tibesti and Ahaggar massifs. Extensive drainage systems, now partly buried by active dune fields, were traced far out into the present desert. Grove (1969) obtained similar results from a study of the Kalahari which was stimulated in the first instance by an examination of aerial photographs.

Unequivocal palaeo-climatological findings such as these clearly implied similar fluctuations in other parts of the tropics currently more humid. They raised the problem once more of their relationship to glacial fluctuations in high latitudes. However, the impact that these shifts might have had on landscape systems in the humid tropics was more difficult to ascertain since the climatic change involved was one of degree, and it was not clear what any particular set of environmental controls might imply for the geometry of landscape if allowed sufficient time. To a considerable extent this is still an open question.

THE IMPACT OF WORLD WAR II

AN ASSESSMENT OF THE FIRST TWO POST-WAR DECADES

Taken as a whole, it is undoubtedly true that the character of geomorphology in the post-War period was changed radically from that before, but the precise impact of the War is still difficult to assess. Post-War military strategies, dictated in part by East-West tensions and the Cold War, certainly generated and funded a substantial quantity of research on terrain analysis, taking this phrase in its broadest terms. The quantification of morphology and the subsequent shift to the measurement of surface processes was probably an inevitable consequence of this available financing. Perhaps the most important effect of the War, at the level of the individual, was to direct people into fields which remained their permanent specialism after the War. In addition to the physical events and experiences of the wartime, the War also marked the end of the Depression years with the hiatus these had put upon budgets and ambitions, both personal and public. In consequence the late forties and the early fifties are placed in contrast with the twenties; a period in the United States when the limited number of academic departments with geomorphologists where still under the influence of Davis's very active retirement and the quasi-dictatorial position exerted by Douglas Johnson, Davis's most noted disciple. Johnson's hold was released with his death in 1944 and at least one doctoral thesis at Columbia was brought off the shelf: it's 'unsatisfactory' conclusions in Johnson's eyes now more suited to a new era.

In the United States the post-War years brought the Horton paper, the rise of the Strahler school and the resurgence of physiographic interest in the Survey. The period saw a return to analysis based on the principles of general physics, and Gilbert was adopted eventually as the intellectual patron saint of this revived philosophy. Such an approach necessarily brought with it the need for process oriented field work and the theoretical, mathematical and statistical apparatus to support it. However, it did not have a significant impact in the Universities until their student base expanded during the 1960s.

The same apparatus led to increasingly refined methods for the objective analysis of landform, a trend paralleled, in different ways and for different reasons, by developments in the United Kingdom, Australia, Canada, the U.S.S.R. and other parts of the world. In Australia and Canada the primary need was still the scientific exploration of their vast, sparsely inhabited terrains. The results, therefore, can be seen as a type of descriptive, scientific, regional physiography. In tropical countries the mandate was similar although the coverage by topographic mapping was still very incomplete. In the United Kingdom the incentive was still precise scientific description, but the items of scrutiny were particular hillslope profiles and valley sides; a change of scale by orders of magnitude and one conditioned by the demands of purely academic interests.

The impact of the new fluvialism was scarcely felt in the United Kingdom until the sixties. The prime reason for this was probably the stranglehold that Wooldridge, based at King's College, London, placed on the system. Wooldridge, as co-author of the pre-War classic on the denudation chronology of South-East England, senior author of the standard British geomorphology textbook, and the only geographer who was a Fellow of the Royal Society (1959, and on the basis of his pre-War geological work), wielded enormous power. His writings, always strongly opinionated, verged latterly on the paranoid. The flavour can be judged from the following extracts (Wooldridge 1958):

> It is quite fundamental that Geomorphology is primarily concerned with the interpretation of morphology, not the study of processes. The latter can be left to Physical Geology.
> No esoteric research in fluid mechanics seems likely to add much to our comprehension of what in essence is simple process in this, and like cases, of developing landforms. [Speaking of river meanders]
> I am not willing to limit my attention and interest to the 'functional significance' of landforms. To such replacement of full binocular vision by an ugly monocular squint, I can only say, as Lord Atlee recently said of television "I don't want it. I don't like it and I won't have it."

In the same paper he speaks of the 'periglacial extremists' in reference to Peltier (1950), and of the 'morphometric squad' led by Strahler at Columbia. It is clear that with such dogma in cold print, the probability that an incoming graduate student would contemplate research in such fields was close to zero and Dury (1982) has recorded that:

> He could be, and often was, as forbidding in demeanour as he was impressive in presence, and it would have been a temarious young researcher who would have dared to challenge him on his own ground.

Only those workers with an independent power base could expect to flourish, and these were few and far between in Britain in the fifties. The group at Cambridge, with particular interests in coastal and glacial processes and landforms under Steers and Lewis, were grudgingly acknowledged by Wooldridge. Linton, at Sheffield, Wooldridge's pre-War co-author, encouraged the developing school of morphologists and slope analysts: a field

Wooldridge could tolerate in view of the fuel it might supply for the Penck versus Davis slope retreat controversy. It is not surprising in the circumstances that Culling, whose work was rejected by Wooldridge, should be publishing in American journals and from private British addresses.

American ideas were also imported through the Cambridge workers, especially Chorley, originally from Oxford, but who spent some time at Brown University in the late fifties and returned to a post at Cambridge, made permanent with the death of Lewis in 1961. Sparks, in his 1960 textbook 'Geomorphology,' made tentative steps to incorporate the American fluvial work into his section on rivers but it was Dury (1959), in 'The Face of the Earth' who did most to educate the British audience about the new American fluvialism. The book appeared as a Penguin paperback and deservedly became a bestseller running into several editions. It is still worth reading and, in 1959, it came to British geomorphology as a breath of fresh air upon a jaded flower.

In retrospect, and with exceptions noted above, it would be easy to dismiss British work during the fifties as being as barren as some of the upland plains under investigation, and American fluvialism as being unduly naive. However, that would be a very unfair assessment. British work was, taken on balance, surprisingly diverse and despite an undue emphasis on denudation chronology, there was considerable attention paid to the operation of contemporary processes as well as the quantitative representation of morphology. It was notable too for a thorough knowledge of what Leopold et al. (1964) termed 'the field problem.' In consequence, it was well-placed, both practically and intellectually, to take advantage of the new philosophies and technologies provided by American work in the fifties. This work was, perhaps, unduly reliant on simplistic concepts from systems theory and statistical mechanics. Nevertheless, it too was well-founded in field work, and Nature, as the nineteenth century discovered, has a way of revealing her complexities slowly, so that a gradual maturation was undoubtedly underway by the time that the death of Wooldridge in 1963 released a surge of British workers onto the floodplains of the Old World.

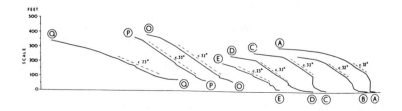

Chapter Eleven

STUDIES OF LANDFORM PROCESSES AND LANDSCAPES SINCE THE 1960s

> *Davis's great mistake was the assumption that we know the processes involved in the development of landforms. We don't; and until we do we shall be ignorant of the general course of their development.*
>
> (J. Leighly, 1940)

> *Contemporary records of geomorphological processes are not likely to represent long-term behaviour sufficiently well to provide any firm basis for understanding landscape evolution.*
>
> (M. Church, 1980)

It is extremely difficult to keep entirely separate the study of landscape processes from the landforms they create. Indeed, such a split, strictly observed, would be contrary to the whole purpose of geomorphological enquiry. However, over the last two decades the emphasis on process has been such that for the purposes of discussion two separate headings are possible. Nevertheless, aspects of one will inevitably leak into the other, and I have provided a section on conceptual and theoretical advances to act as a bridge both between sections, and between what may be, and what ought to be, converging schools of thought.

THE SHIFT TO PROCESS STUDIES

General and institutional indicators

The switch to studies of landform processes from those of regional landscapes was accomplished within a few years in the United Kingdom. It had already been accomplished in the United States, but was much more muted on the continent, where interest in processes accompanied a prevailing interest in climatic geomorphology with its inevitable focus on regional scale landforms. In tropical countries the response was a lagged echo from European interests.

The early 1960s make a convenient benchmark from which to examine present geomorphology because any effects from the War were well past, economies were expanding, and the wartime 'baby bulge' brought an influx of students from 1960 onwards with a corresponding expansion in University posts, and later in the sixties, increased numbers of graduate students and a wider variety of research opportunities. Other factors also favoured a shift to process work. It savoured of a more scientific approach, an important point wherever geomorphology was institutionally lodged in Geography departments. Geographers, enmeshed in the Quantitative

Revolution, were anxious to display scientific orthodoxy, and themselves were rejecting a regionally scaled, descriptive paradigm for their subject. Process work, inevitably linked to restricted areas and specific experiments, had the aura of an applied subject, of being concerned with the operations of the real world. The roots of this work in the USGS had been exactly that; flood control problems. Work of this character, as well as being a natural response to the rejection of Davisianism, more easily attracted grants from funding agencies, and was seen to be more readily justified.

The beginning of the 'modern' period was also marked, in Britain at least, by the establishment (1961) of the British Geomorphological Research Group (BGRG) under the leadership of Linton. It was originally an independent entity but is now a Research Group of the Institute of British Geographers. The BGRG began publishing Technical Bulletins in 1969 through Geo Abstracts and later (1976) instituted its own journal 'Earth Surface Processes,' a title subsequently expanded (1981) to include 'Landforms.' A corresponding research group did not emerge in the United States until 1979 when a Special Interest Group of the Association of American Geographers was formed, although the annual Binghamton Symposia and the biannual Guelph Symposia started in 1970 and 1969 respectively have provided rallying points for North American geomorphologists. Another indicator of the new mood, and the expanding literature on the subject, was the initiation of Geomorphological Abstracts in 1960 by Keith Clayton, an enterprise that has now burgeoned to Geo Abstracts, with ancillary publishing interests in the earth sciences.

Practical applications of geomorphology have seen an increasing emphasis during the 1970s with the publication of Cooke and Doornkamp's 'Geomorphology in Environmental Management' (1974), the volume on applied geomorphology recording the contributions of a Binghamton Symposium (Craig and Craft 1982), and a book of essays (Hails ed. 1977). Practical and applied geomorphology has been an abiding concern in Eastern Europe and the Soviet Union, where there has been a considerable development of cartographic methods for the depiction of surface characteristics (Demek 1972).

An index of maturity is provided by the growing interest in the history of the subject. The definitive volumes of Chorley, Beckinsale and Dunn (1964, 1973) are the first two of four that have been promised. British geomorphology has been well served by the masterly survey of Davies (1969) which covers the period 1578 to 1878 and which, because of the central position of Britain in pre-twentieth century geology, provides a more comprehensive coverage of the subject's history than might be suspected from the title.

As a result of all the work on processes there has been a flood of textbooks entitled, or justifying themselves in terms of, process geomorphology. These run the gamut from

specialist texts of particular topics: process mechanics
(Carson 1971), slopes (Carson and Kirkby 1972), deserts,
(Cooke and Warren 1973), drainage basins (Gregory and Walling
1973), glaciated landscapes (Sugden and John 1976), karst
(Sweeting 1972), beaches (Komar 1976) and weathering (Ollier
1969) to comprehensive textbooks (Ruhe 1975, Ritter 1978,
Embleton and Thornes 1979, Derbyshire et al. 1979). Clearly,
the subject has not been badly served.

Any attempt to assess contemporary work, which is still
growing dynamically, can be made only by inspecting current
perspectives of the results achieved, and the proposed
direction of future work. All such assessments are involved
inextricably with the work which they try to assess, and may
turn out to have a limited historical validity.

Fluvial geomorphology

To retain an interest in rivers while rejecting the Davisian
Cycle required some similar, but alternative focus. The
drainage basin became that focus, not at the regional scale,
but at the specific order scales derived from the Horton/
Strahler system of network ordering. The text by Leopold,
Wolman and Miller (1964), summarising a decade of work,
became the 'bible' of the new orthodoxy and devoted a chapter
to the drainage basin as a unit of study. The term 'first
order basin' was soon adopted as a catchword amongst
geomorphologists and typified the new interest in small scale
systems, although, as Gregory and Walling (1974) note the
spatial scale was Hobson's choice. The drainage basin theme
was expanded by Gregory and Walling (1973) in their book
'Drainage Basin Form and Process.' In addition, the river
channel itself, whose behaviour had been taken as understood
before 1950, became the focal point of morphological
interest. The movement of the channel would, in time, adjust
to changing system inputs, create meanders, braids,
floodplains, terraces and valley side bluffs. Ultimately, a
thorough understanding of channel behaviour ought to
illuminate the perennial problem of valley formation, another
topic long taken for granted. However, concerns with valley
side slopes, the other component of river basins, were not
completely neglected. The Sheffield school of morphologists
was already moving into the formal study of slope profiles
and processes (Young 1960, 1961, Bunting 1961), and an
integrated view of basin behaviour became the norm.

A primary interest of all parties was the measurement of
the rate of sediment loss, both solid and soluble, from
slopes and through channels, with an emphasis on attempting
to describe the spatial variation in the loss rate. Britain
lacked the long term records of river flow and sediment
discharge that were available in the United States. Thus the
Vigil Network system (Leopold 1962, Slaymaker and Chorley

1964) with its emphasis on small, easily and inexpensively instrumented basins attracted British attention, and this was reinforced when the period 1965-1974 was adopted as the 'International Hydrological Decade.' A UNESCO resolution (2323) in 1974 established an International Hydrological Programme as a follow up to the IHD. Renewed interest in world erosion rates, and the corollary, rates of vertical movement of the crust, acted as a general framework for local studies (Williams and Meade 1983, Meade 1969, Schumm 1963, Fournier 1960).

One realization from this work was that regional erosion rates likely are enhanced, with respect to the geologic past, by the environmental depredations of man, added to the fact that many high latitude terrains are formed of readily eroded glacial deposits. In addition, our knowledge of the processes operating in tropical regions, and their rates of operation is still very scanty. Douglas (1978) noted that 'Tropical geomorphology is slowly emerging from a somewhat dilettante occupation for expatriate, peripatetic, physical geographers and Quaternary geologists.' Thomas (1978), on the other hand, remarked in the same volume that 'outside Australia, however, committed research in tropical and subtropical environments is if anything dwindling.' This particular problem is political. The two decades following the War led up to the time of Independence in many African countries, and a growing assertion of nationalism gradually displaced expatriate faculty in Universities and Government agencies. Although many African nationals have trained in Australia, Europe, or the United States, research funds are often extremely limited and political instability has taken its toll in discouraging field work. The two decades that saw the greatest expatriate involvement coincided, by and large, with the period before the shift to process studies.

What resulted from over a decade of process work? A British assessment of fluvial research showed a marked increase in significant indicators (Gregory 1978). In the period 1965 to 1970 the Natural Environment Research Council (NERC) funded work on 47 catchments in Britain, river gauging stations increased from 200 in 1960 to 1200 in January 1975, and while the fluvial entries in the BGRG's Register of Research rose only 5% to 25% from 1963 to 1975 it is safe to assume that the character of this work had changed radically, from a regional to a process orientation. Yet, in hard terms, what had been achieved was the accumulation of many useful results, the most general of which was that the dissolved load of rivers in non-limestone areas, contributes a far more significant proportion to erosional losses from basins than was previously thought (Walling 1977). However, what was still lacking was any sort of synthetic meaning connecting the operation of the system at different spatial scales. In a volume devoted to 'River Channel Changes' (Gregory ed. 1977a)

Thornes (1977) drew attention to three points: 1) the need to assess channel characteristics in terms of the whole network, 2) a need to consider unstable, as opposed to stable, systems, and 3) a need to change the spatial scale of concern back to that of the whole basin. This, certainly, was potentially more productive than calling for more of the same.

To some extent this more integrated view was supplied by Schumm (1977) in his book 'The Fluvial System,' a book which stemmed from an invitation to provide a summary view of the workings of the river basin for economic geologists. It is not without significance that this imperative required an approach both from the point of view of contemporary processes, and from an historical perspective that could throw light on the distribution of economic deposits in ancient and modern alluvial systems. The emphasis of the book is sedimentological; the episodic transit of sediment through the system during intermediate spans of time, i.e. neither cyclic time, nor spans measured in a few days or years. It is not possible yet to assess the impact that Schumm's book ultimately will have on fluvial geomorphology, but it is already cited widely, and in recent years there has been a definite shift to problems with intermediate timescales (Thorn ed. 1982, Cullingford et al. 1980, Brunsden and Thornes 1979).

On the complementary subject of slopes, a review by Young (1978) of the period 1970 to 1975, saw it as one in which slope studies held their own, but did not match the expansion period of the 1960s. Predictably, mild growth was seen in the fields of erosional processes and their rates of operation, instrumentation, landsliding and the iterative modelling of slope profiles. Decline was seen in the realms of theoretical slope forms, and in formal studies of slope evolution. The field was admirably summarized in volumes by Selby (1970), Young (1972) and Carson and Kirkby (1972), which indicates that already a degree of coherence had been achieved by the early 1970s, and from which additional progress might require another decade or more of accumulated field work. The 1985 Binghamton Symposium is scheduled to be upon 'Slopes' so that a new summary of the field should be available then. By 1974 Young could report that rates of slope retreat had been confirmed, from a wide variety of environments, to be essentially the orders of magnitude reported by Schumm (1963) a decade earlier: further work on this topic would appear to be redundant. Even so, Young still foresaw the need for more work on the mechanics of slopes and processes (and by 1978 Prior admitted that many of the contributions on mass movement were by civil engineers), an emphasis on the soil catena, and a renewed use of slope studies in regional and historical studies as an aid in debates about landscape evolution over cyclic time.

The related topic of weathering has seen steady progress. Yatsu's book on 'Rock control in Geomorphology' (1966) did not have the impact that it deserved but experimental studies in weathering seem to have progressed from the period when Sparks (1971) reported Williams' remark that during experiments on the freeze-thaw mechanism in chalk lithologies the apparatus broke more often than the rocks. Not unexpectedly, weathering processes are turning out to be more complex than earlier literature suggested. The importance of salt crystallization in a wide variety of environments, including very arid ones, has been appreciated. A recent contribution recognizes its importance in association with freezing (Williams and Robinson 1981), which may be relevant in explaining wide raised platforms on resistant lithologies in western Scotland.

Old dogmas are difficult to displace and it is only recently that the basic mechanism of the freeze/thaw process has been called into question (Ives 1973, White 1976, Thorn 1979, Lautridon and Ozouf 1982). It now seems likely that a variety of inter-related processes are involved, several of which may not even involve freezing at all, e.g. the molecular phenomenon of ordered water growth (Dunn and Hudec 1966), although most of them involve temperature cycling around the freezing point.

Weathering, the breakdown of solid rock to smaller particles or solution products, lies at the very heart of geomorphology yet the subject has still failed, in large part, to make any convincing connection between what is either a molecular or very small scale process (in spatial terms) and the development of regional landscapes. Weathering is to cyclic geomorphology as particle physics is to cosmology and it faces similar problems in trying to make a satisfactory connection between the two.

Glacial geomorphology

Although glacial geomorphology was not bound by cyclic arguments (as Hobbs had pointed out long ago glacial 'cycles' were essentially climatic), it was long concerned predominantly with the mapping of regional glacial landforms from drift stratigraphy in an effort to delimit the extent of Quaternary ice. The approach was primarily geological and reflected the institutional association of geomorphology with geology in the United States. The interpretation of existing landforms on the basis of actual glacial processes, when it was done, was done primarily on the basis of analogues and rested on an over-simple view of glacial sedimentation. A particular problem here was that it was much easier to study glacial behaviour at existing ice margins, than the glacier bed hundreds of kilometres behind the ice front. Consequently, the emphasis of work until well into the 1960s was upon the progress of deglaciation, partly because of this marginal bias, and partly because the evidence of glacial retreat is, for obvious reasons, the best preserved. Yet, at the height of glaciation, ice covered some 30% of the earth's surface and much erosion and deposition must have occurred which cannot be attributed to the processes of deglaciation.

A revolution in glacial geomorphology was achieved in the early 1970s when Boulton (1972, 1974), from the basis of glacial physics, suggested that the sole of a glacier might comprise four zones each with a different thermal regime, and each likely to give rise to a different suite of erosional and depositional landforms. These suggestions heralded the birth of modern glacial sedimentology, although the roots of this work can be found in the books of James Geikie (see Chapter 7) and the neglected work of Carruthers (1953). From a rational basis in sedimentology, an understanding of the constructional landforms of glacial deposition might be achieved. For this approach to be practical at the scale of a continental ice sheet clearly it was necessary to be able to predict the thermal properties at the base of a formerly existing ice sheet. Attempts to reconstruct former ice sheets required precise information on marginal limits and their dates, estimated ice depths and ice velocities, and climatological data pertaining to the estimated ice surface. Admirable summaries of existing research on all aspects of glaciation, past and present, were already available in books such as Embleton and King (1968), and Flint (1972). Thus, the sudden shift imposed a new perspective on existing material, and rapid progress was possible. As a result several attempts have been made since the mid-1970s to reconstruct the major Pleistocene ice sheets, of which the most notable are Boulton (1977), Sugden (1977, 1978) and very recently the CLIMAP project which resulted in a comprehensive volume 'The Last Great Ice Sheets' (Denton and Hughes eds 1981). They noted that the attempt to model the growth and decay of these ice sheets dynamically had foundered on a lack of information, and consequently the volume limited itself to static reconstructions at assumed points of steady state in ice sheet size, and to modelling dynamically limited aspects of ice flow. Even so, it does presage the future since it connects the fields of glacial physics and glacial sedimentology; and the dynamical modelling of ice sheets will be the next step forward. The application of the thermal regime model still has to overcome the severe problems posed by the advance and retreat of a single ice sheet, the probable failure of the sheet to achieve a steady state morphology, and by multiple glaciations within the Quaternary.

Acceptance of the thermal regime paradigm in glacial geomorphology is currently strongest in Britain and Scandinavia, with the United States and Canada only slowly and selectively developing the new methodology, whilst other European countries including Russia have yet to accept it. The resistance to acceptance is partly founded on the fact that the new interpretations negate much previous work in drift stratigraphy, which is in turn rooted in a classical geological approach to drift sections.

A complementary approach to whole ice sheet modelling has been the attempt to understand the ability of large ice sheets to deform the earth's crust isostatically. The late 1950s and the 1960s were years of intensive activity in the Canadian arctic, which were partly stimulated by the defensive needs of North America, and the result has been an exponential explosion in the amount of material available on all aspects of high latitude landscapes, ancient and modern (Parry 1967). One facet of this work has been the mapping of isostatically deformed late-glacial and postglacial shorelines in the Canadian arctic. The delimitation and dating of these shorelines is crucial to establishing the pattern of deglaciation for the Laurentide ice sheet, and this in turn provides valuable inputs to geophysicists anxious to understand the rheological properties of the earth. A masterly summary of this work appeared in Andrews (1970), and the methodology has implications for Highland Britain and Scandinavia. It had taken a century from the first recognition of the problem to a full statement of its geomorphological effects.

Coastal geomorphology

By the early 1960s the analysis of processes on soft rock coasts was already well underway, and King's 'Beaches and Coasts' acted as a signpost for process work in the future. Consequently there have been fewer radical changes in thought in coastal geomorphology than in other branches of the subject, and a recent review (Walker 1978) merely remarked on a continued emphasis upon process studies, although King (1978) noted the increasingly sophisticated technology in use. Sediment movement in the beach and wave environment still presents many puzzles, even to the extent of establishing the downward limit of wave and tidal action. An interesting development has been the use of standing waves, edge waves and cellular circulation systems with their attendant instability mechanisms to build three-dimensional process models for the explanation of rhythmic features of the coastal plan, such as beach cusps and regular bays.

Given the importance of the breaking wave to coastal processes, a classification of coastal environments by tidal and wave environments by Davies (1964, 1973) has been widely used, and there has been continued concern with the problem of changing sea levels during the Quaternary. Fairbridge (1961) and Shepard (1963) both published curves of the recovery of sea level to its present value after the melting of the ice sheets. The susceptibility of coasts to slight shifts in the mean level suggests that the establishment of the trend of sea level is critical to short and long term planning. This concern brings coastal geomorphologists into contact with quaternary specialists modelling the decay of Pleistocene ice (Clark and Lingle 1979) because the coast is part of a tectonic environment that interacts with the changing level, being deformed isostatically by the additional water load placed upon it as sea level rises (Bloom 1963). In this connection a classification of coasts in terms of their relation to plate boundaries will have meaning for temporal trends of change (Inman and Nordstrom 1971), and the interpretation of coastal morphology over cyclic time.

The expansion of interest in the arctic, noted under the glacial heading, has also caused an expansion in studies of arctic coasts. The presence of ice for much of the year leads to the protection of the coast from many of the normal effects of the wave environment, but creates its own landforms from the effects of ice floes impacting on the shore. Because longshore processes may be reduced by more than an order of magnitude from what would be expected on a temperate coastline of otherwise similar geology, arctic coasts tend not to display smoothing of the coastal plan.

The study of coastal processes is pre-eminently an applied field because of man's predilection to coastal domiciles. However, Walker (1978) has sounded the cautionary note, not without value to other parts of geomorphology, that such applied work is not necessarily best suited to answering fundamental questions, even in the process field, and the consequences of underfunding non-applied research during a time of recession must be emphasized.

CONCEPTUAL AND THEORETICAL ADVANCES ACROSS THE SUBJECT

Dynamic equilibrium

As a result of the work on rivers in the United States during the 1950s the prevailing philosophy brought to geomorphology in the 1960s was that of systems theory and its concern with stable, though not static, equilibrium states. Chorley (1962) summarized this thinking and used it to contrast the supposed closed system of classical Davisian theory with the open systems which characterized channel hydraulics. The intellectual origins of this approach, in geology, can be traced to Du Buat (1779), Gilbert (1877) and more recently Mackin (1948), in a famous paper concerned with the vexed question of river 'grade.' Mackin explained that a river would maintain a steep slope in 'bedrock' of cream cheese if it was necessary to transport coarse bedload! However, while Mackin made the first post-War statement in geology about open systems with negative feedback (but see p163 for Leighly's neglected paper of 1932), his philosophy did not, in general, mesh with the statistical approaches to equilibria typical of the Leopold group, as their contrasting contributions in Albritton (1963) document.

Space, time and causality

While the emphasis remained on short term process studies, a classical open systems philosophy was of value, but Schumm and Lichty (1965) found it necessary to point out that the same behaviour might look very different when viewed over different temporal scales. They reconciled the timeless and timebound aspects of geomorphology by noting that 'the distinction between cause and effect among geomorphic variables varies with the size of the landscape and with time.' Therefore, for example, the width of a channel cross-section is essentially a fixed independent variable for the passage of a single flood flow taking a few hours or days. But, over a sufficiently long period of time the channel width can adapt to a changed flow regime, and so at periods of, say, tens of years or more, the width becomes dependent on other aspects of the system. In identifying the main elements of landscape study as either dependent or independent, over varying timescales, 'the disparate points of view of the historically oriented geomorphologist and the student of process can be reconciled.' The authors recognized three types of temporal scale: cyclic, graded and steady, in decreasing length, measured in years; the last appropriate to process work, the first to Davisian time, and the middle one to time scales between, in which gradual, but small, changes in the mean state can be detected; appropriate for example, to timescales such as the postglacial.

I noted the impact of the magnitude/frequency concept in the last chapter. Together with the paper by Schumm and Lichty it has provided the main conceptual basis of thinking up to the present, placed as it was in the general context of systems theory (Chorley and Kennedy 1971, Andrews 1975).

Thresholds, complex response and Catastrophe Theory

As field work progressed, and more elaborate problems were investigated, the need for a broader theoretical basis became evident. The first indication of the new thinking came when Schumm (1973) noted that many processes were most easily understood in terms of thresholds. Below the threshold little happened, but above it the system in question moved to another equilibrium state, and the overall behaviour was radically different.

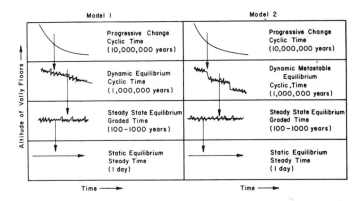

Figure 16. *Models of landscape evolution. (a) Model 1: equilibrium components of Davis model of progressive denudation (Figure 1-2) (b) Model 2: equilibrium components of model based on episodic erosion. (from Schumm 1977).*

Schumm indicated that this was evidenced by the fact that a simple change of baselevel, in a laboratory stream system, apparently produced a whole series of changes, a result Schumm termed 'complex response.' The idea of thresholds was taken as the basis of the 1976 Binghamton Symposium, and the basis was broadened by inviting other contributors from across the subject before the volume was published (Coates and Vitek 1980). A full statement of Schumm's views can to be found in Schumm (1977, 1979) where he blends the threshold idea with that of equilibrium and suggests that geomorphological systems display 'dynamic metastable equilibria.' In such a system fluctuations may occur about a mean which changes only slightly in graded time, but which on encountering a threshold will shift suddenly to a different equilibrium state.

In the introduction to their volume, Coates and Vitek noted a 'nomenclature gap' which they thought was inhibiting discussion (vide Davis' remark about antecedence!, p149), and they noted too 'a lack of universal paradigms' amongst geomorphologists. They offered the notion of 'threshold' as a significant one since it integrated the notions of steady state equilibria, and the critical notion of systems undergoing a transition from one state to another. This idea was already being popularized during the 1970s elsewhere in science in the more elaborate form of 'Catastrophe Theory' (an idea that has nothing to do with catastrophism in nineteenth-century geology), and which may be better understood as a method of analyzing all possible types of system behaviour under a limited number of independent controls.

Catastrophe Theory has had such a bad press in terms of its actual applications, especially to social science, that it has not caught on in the way one might have expected given its apparent applicability to the problems in hand, although discussions of fluvial geomorphology using Catastrophe Theory have been given by Graf (1979b) and Thornes (1980).

All of these shifts can be interpreted as attempts to stay within the constraints of systems theory while adjusting to more complex field problems, considered on more extended spatial and temporal scales than the river bed, day by day. Complete applicability may be very difficult to accomplish since Ford (1980) observed that 'discrete thresholds are difficult to find in many parts of the karst system, but certain limits are clearly defined.' Thresholds may appear even in the karst system if the temporal scale is extended, but that remains to be tested.

Neocatastrophism

Nineteenth-century geomorphology was concerned with over-throwing all vestiges of catastrophism as a formative agent of landforms, in favour of the scientific respectability attaching to the principle of uniformity. The twentieth century has seen a gradual reversion to the recognition that surface processes may not be infinitesimally smooth, even when viewed over graded and cyclic time. The derivatives of equilibrium theory described above are attempts to accomodate thinking to a more discontinuous reality even if it is less in keeping with the classical mathematics commonly used in quantitative modelling.

Neocatastrophism (Dury 1979) is an overt recognition of the potential significance of 'large,' and possibly unique, process events to features in the geological record. As a term and concept it has 'escaped' from palaeontology, where it was involved with mass extinctions, into geoscience and in geomorphology it may have most significance on interfluves, rather than in channels where more frequent events may reshape landforms, or where sediment protects the bedrock form of the landscape. However, Patton and Baker (1978) have shown that very rare events, in terms of basin hydrology, have had very significant effects on the morphology of channels in Texas. Nineteenth-century catastrophism was essentially magical, but neocatastrophism, even though it depends on known and understood events, has yet to be assimilated formally within a greatly expanded uniformitarian umbrella. It is not greatly different from the mild catastrophism with which Lyell modulated his uniformity (see Chapter Five) and its re-emergence may indicate a need to reconstruct the magnitude/frequency concept, which despite, or perhaps because of, its general acceptance has not been as closely examined as it might have been.

Continental workers also have been sensitive to the need to incorporate extreme and catastrophic events within the conventional rubric of morphogenetic agents. Tricart and Cailleux (1965), Starkel (1976) and Budel (1982) all mention these two terms although only Starkel discusses them in terms of the Wolman and Miller paper (1960) and unfortunately he fails to distinguish clearly between the terms 'extreme' and 'catastrophic.' In all cases the emphasis is placed upon the fact that the events in question have high recurrence intervals and yet, in most cases, are sufficiently frequent to be significant agents in shaping slope form and river channels. There seems, therefore, to be a distinct difference between this emphasis, under which the events are extreme elements of the normal climatic regime, and neocatastrophism proper in which the events are of truly massive magnitude, and are genuinely ex cathedra.

Thus the issue returns to the point made: that the whole conceptual framework needs a thorough re-thinking, especially since, as Starkel remarks, it is important to distinguish between extreme events in the meteorological record, and those which are extreme in their impact on the landscape. Gardner (1977), for example, discusses a catastrophic flood on the Grand River in Ontario with a recurrence interval which may have approached 500 years, and yet which had minimal geomorphic impact.

New modes of diastrophism - plate tectonics and neotectonics

The 1960s was a period of revolution in geology during which the long out-moded theory of Continental Drift was resurrected and re-vitalized as the prevailing theory of plate tectonics. Hallam (1973) has documented the history of thought on the topic. Although geomorphologists are well aware of the theory there has been no strong movement to accomodate thinking to the new diastrophic ideas. Melhorn and Edgar (1975) considered the issue at length for the specific case of eastern North America but their conclusions merely re-affirmed Schumm's general conclusion of 1963: that vertical movements were restricted in time and that there was ample time in between for any of a variety of planation mechanisms to have produced widespread surface levelling.

Others who have written regional studies within a framework of plate tectonics are Judson (1975), Brown (1979) and Ollier (1979). The last two are discussed further below, but there still remains the need for a formal study of the relevance of new diastrophism to landform evolution on long timescales. The focus on the short timescales pertinent to process studies mainly has been responsible for this neglect. In the long term the deficit must be rectified since the motion of plates has the potential to integrate a variety events previously considered, in Davisian terminology, as 'accidents.' Modes of vertical motion, long term climatic change through the latitudinal drift of plates, variations in continentality, the onset of Ice Ages (Fairbridge 1973), even the appearance of volcanism: all seem likely to emerge as inevitable consequences of plate motion.

At smaller spatial and temporal scales the topic of neotectonics (Obruchev 1948): the motion of the earth's surface since the Miocene, with an emphasis on vertical changes, has attracted some geomorphic interest since whether it be in regions of isostatic uplift, or those affected by orogenic processes there are some regions, as Schumm showed in 1963, where uplift exceeds denudation rates by an order of magnitude and where the geomorphologist cannot fail to adjust his thinking to the rate of uplift. The development of neotectonics by Russian geologists partly reflects their aversion to the theory of plate tectonics (Beloussov 1970) and partly the environmental bias created by the planation surfaces of a continental interior where movements since the late Tertiary have been predominantly vertical.

The 1984 Binghamton symposium is to take tectonic geomorphology as its theme, and it may be that a new and comprehensive perspective of this central problem in geomorphology will emerge from that meeting.

Theoretical and quantitative developments

Chorley (1978) remarked that at the mention of 'theory' geomorphologists reach for their soil augers. The thinking discussed above is primarily conceptual, not theoretical, if a strict approach to the term 'theory' is adopted (Thornes 1978). However it is interpreted, theory, which is after all deductive in nature, has not been prominent since 1960, and perhaps this is an allergic reaction to the supposed horrors of Davisian Cycle. The explicit work in theoretical geomorphology has been that of Scheidegger (1962, 2nd ed. 1970), but its reception has been muted, and it is not widely cited. The reasons probably lie in its recondite mathematics (to the average geomorphologist) and its otherwise eclectic approach to individual landforms, many of which are apparently chosen for their susceptibility to analytical treatment. It is more a theory of landforms than a theory of geomorphology; it offers no disciplinary synthesis.

More active has been the numerical modelling of certain categories of landform. In coastal geomorphology the best example is the spit simulation program of King and McCullagh (1971), while in the fluvial realm drainage basin simulation began in the late 1950s but has now graduated through two dimensional slope modelling (Young 1963, Carson and Kirkby 1972) to the three-dimensional simulation of drainage basin evolution (Vanderpool 1982, Ahnert 1976, and Sprunt 1972). The most significant element in this work is the use of the continuity equation at a point on the slope profile to represent soil and surface processes. Integration, analytical or numerical, yields the shape of the slope profile as a function of time and the 'characteristic form' of the slope can then be studied (Carson and Kirkby 1972). The numerical simulation of continental ice sheets, both in steady state, and in growth and decay, is a central feature of contemporary glacial geomorphology. All of these approaches depend on the computational and graphical facilities of modern computers.

The morphometric analysis of landscape, and especially the measurement of landform attributes from topographic maps, has seen continual activity throughout the period and the work has proved difficult, though not impossible, to relate to applied research in river basins on the one hand (Gregory 1977b, Patton and Baker 1976) and the evolution of landscape over cyclic, or even graded, time on the other. Present work promises to throw some light on the relationship of the internal structure of the drainage net to the gross aspects of geological structure, as a basis for determining the effect of structure on network configuration (Flint 1980).

Geomorphologists have not been slow to utilize the quantitative methods made readily available in program packages for mainframe computers since the mid-1960s, but, with few exceptions, the subject has not been swamped with quantitative methodology to the exclusion of useful results. The demands of Strahler's methodology strained hand computation and punch-card techniques to the limit. Chorley (1966, but written in 1961) provides a useful summary of the quantitative methods of the pre-computer era. Gregory and Brown (1966), on the other hand, acted was clarion call to the computer age. However, apart from studies centred on morphometry, with data collected mainly from maps, the acquisition of large amounts of consistent data is not easy for the field geomorphologist, and computer centres have not been overloaded with field analyses. Doornkamp and King's mistitled 'Numerical Methods in Geomorphology' (1972) summarized standard statistical methods, elementary analysis and rudimentary simulation, but it did not address the more important general problem of articulating theory in a numerical fashion.

In general, the development of theory has been confined to individual strands of thought on specific topics while the subject as a whole has been held together by concepts loosely drawn from systems theory and broadened to suit a widening spatio-temporal scale.

CINDERELLA IN THE KITCHEN? LANDSCAPE STUDIES SINCE 1960

The period since 1960 has been widely regarded as one in which, due to the paramount interest in process, the study of landforms has been neglected. In reality this is very far from the truth. The scale of spatial and temporal interest has certainly shifted from the predominantly regional to the myopically local, but there are signs now that the scale is increasing again, and there has always been significant residual attention paid even to regional landscapes.

Denudation chronology in Britain appeared to suffer a fatal blow with the death of Wooldridge (1963), but it was primarily symbolic: Dury, Chorley and Vaughan Lewis were beginning to preach and practise the importance of measuring processes in coastal and glacial geomorphology, and in karst systems the practice of measuring the solution load of underground water was already underway. The explosion in student numbers also would have weakened the grip that a few individuals could exert on the entire system.

In glacial and coastal geomorphology there is a natural tendency to focus on the temporal scale provided by the later stages of the last ice advance, and on the postglacial, a period also of interest to archaeologists and Quaternary botanists. The use of radio-carbon dating since 1950 has meant that events within the last 40,000 years can be pin-pointed with an accuracy previously unknown, and is an additional reason for the considerable attention paid to late Quaternary chronologies since the early 1950s. A problem with this methodology is that it tends to reinforce a stratigraphic approach to regional mapping in which considerable detail at a single section fails to compensate for inadequate spatial controls on the dating.

The focus on process discussed above, and its close relation to applied geomorphology, has caused consideration to be given to man's effect in creating, or destroying, landforms. This problem, considered in postglacial time, has been reflected in valuable studies in the Mediterranean (Vita-Finzi 1969) and the Near-East (Davidson 1980), and for the nineteenth century in the semi-arid American south-west (Cooke and Reeves 1976).

I intend now to discuss a number of topics to which landforms are central, and which have been of persistent interest in the last two decades, despite the focus on processes.

The valley meander problem and misfit streams

W.M.Davis was probably the first to use the term 'underfit' (although Hall (1815) recognized the condition) to characterize streams whose present meanders are at least an order of magnitude smaller than those of the valley meanders in which they lie. The rivers are insulated from the valley floor and walls by a thick sedimentary infill. His explanation depended on stream captures, climatic change, or the direct discharge of glacial meltwater from ice-dammed lakes into the stream concerned. However, it was Dury (1954) who first identified the condition as regional and widespread and who coined the general term misfit, although most examples are, understandably, underfit.

Sub-surface exploration of the sedimentary fill confirmed that valley bottoms had a similar geometry to regular stream meanders, but of much greater magnitude. It also revealed that the filling process was episodic with gradually reducing channel dimensions. Various methods of dating indicate that the filling began in the very early postglacial and was completed in most areas by 9000BP (before present) and imply that the valley meanders themselves were active in the late glacial, about 12,000BP, although there are local variations (Dury 1964a, 1964b, 1965, 1977).

The problem posed by these channels is that of accounting for the magnitude of the hydrological events that formed them. By analogy to present stream systems Dury relates the formative discharge of valley meanders to flows with a recurrence interval of 1.58 on the annual flood series. By back-tracking this value, which is thought to be related to the channel dimensions through hydraulic geometry, Dury was able to estimate increases in rainfall of the order of 50% to 100% for the limited periods during which the channels were formed. This, he believed, implied an increased 'frequency and power of frontal storms' during the late glacial.

Dury's work takes on added significance in the light of Baker's work on the response of streams in central Texas to catastrophic rainfall and runoff. Patton and Baker (1978) demonstrate that valley morphology is being formed actively by flood events with recurrence intervals of several hundred years, and that even incipient floodplains and bars are reshaped by such discharges. Thus the topic of valley meanders, and their formation, can be linked to the general problem of valley formation in the fluvial system, a subject in which details have usually been taken for granted.

Despite two decades of field work on rivers, the overwhelming predominance of research has been on alluvial channels. With the exception of the work just mentioned only very recently has a concerted attempt been made to examine bedrock channel form and process (Howard and Dolan 1981, Graf 1979a). Likewise, only a limited amount of work has undertaken the physical modelling of valley formation (Shepard and Schumm 1974), as distinct from alluvial systems.

Postglacial dynamics in alluvial systems

Streams which are insulated from bedrock by an alluvial fill can readily adjust to changing system controls. The Mediterranean region is a classic area for studying postglacial alluvial systems because historical documentation

stretches back to the earliest Classical authors, and
archaeology provides an additional source of information for
relative and absolute dating. Vita-Finzi's conclusions from
studying the alluvial structure of Mediterranean valleys is
that there is a broad sychroneity across the entire basin for
a sequence that begins with an older valley fill that was
deposited by about 10,000 years ago. Stream incision into
this fill was then typical throughout the subsequent millenia
until it was replaced by a renewed phase of valley floor
aggradation that began about 500 AD, although the precise
date of onset varied around the Mediterranean basin. The last
few centuries have seen a second period of incision.

It would be tempting to ascribe these changes to unwise
agricultural practices and general deforestation, especially
during the Classical period. Davidson (1980), for example,
thinks that agricultural malpractice may have exacerbated
erosion. However, the general synchronism of the phases
across the entire basin, the inadequacy of a multitude of
local causes, and the knowledge that Europe suffered a
sustained period of much cooler weather, known as the Little
Ice Age, all suggest that a general climatic explanation is
better than one based on a large number of different local
circumstances. Vita-Finzi noted that a shift to the aggrad-
ational phase may hold within it the seeds of subsequent
incision. Rapid aggradation implies soil erosion on the
tributary slopes, and eventually this supply will cease
(weathering limited erosion) and the more resistant clay base
to the soil, or the exposed bare bedrock, will promote
increased surface runoff and trigger incision.

Similar complexities may befog studies of much shorter
time spans. Cooke and Reeves (1974) studied arroyo behaviour
in Arizona and coastal California and found that although a
synchronous phase of entrenchment occupied the period 1865 to
1915, with a peak around the 1880s, only in California could
they confidently ascribe major changes in the floristic
composition to the effects of men and cattle. The vegetation
changes then induced increased runoff and stream incision. In
Arizona, in contrast, 'secular, and short term climatic
changes may have been an important cause of vegetation
changes, and it can be argued that overgrazing may have been
less significant as an ultimate cause of entrenchment than
some have assumed.' Very recently Alford (1982) has
reconsidered the San Vicente arroyo in New Mexico, which
entrenched overnight after a large flood in 1895. This he
relates to documented evidence of severe overgrazing in the
previous twenty five years, rather than to an alternative
model which relies upon exceptional storm rainfall as the
single causative event.

These two examples show that regional studies, in the
context of an understanding of drainage basin processes have
done much to illuminate the 'complex response' for stream

systems identified by Schumm, and their apparent sensitivity to certain subtle changes in surface hydrology illustrates the applicability of the threshold model.

Glacial terrains: implications of the thermal regime model

Boulton's thermal regime model described in an earlier section, when related to a reconstructed ice sheet, makes seminal suggestions about the spatial organization of glacial landforms, even when the complications of non-steady ice sheets and multiple glaciation are taken into account. The most interesting work in this direction is the reconstruction of the Laurentide ice sheet by Sugden (1977, 1978) and its impact on glacial erosion. Sugden's conceptual model is based on Boulton's thermal regimes as recognized in a steady state reconstruction of the ice sheet. One fascinating feature of the model is the reappearance of glacial protection in areas where the ice sheet is debris-free and cold based, i.e. frozen onto the bedrock. The northern slopes of the Laurentide ice sheet appear to have exhibited such behaviour since many areas reveal few signs of recent glacial erosion. However, such protected areas are interspersed with zones of warm-melting and warm-freezing in which considerable areal scouring or selective valley trough erosion may occur. These general effects are locally modified by pre-existing topography and local geology, especially by the distinction between permeable and impermeable beds, and the effect that this may have on the evacuation of subglacial meltwater and the viscosity of the basal till.
    Clearly, the conclusions from this sort of work are still in the exploratory stage, but they are suggesting that glacial geomorphology stands to gain an added regional, and potentially calculable, coherence to augment the specific aspects derived from chronologies of glacial retreat.

Late Quaternary raised shorelines

Shore processes on hard rock coasts tend to produce a shore platform, backed by a cliff, and mantled with scattered deposits usually concentrated towards the upper limit of high tide as a beach of sand or gravel. The short period during which sea level has been close to its present level, about 4000 years, means that on resistant lithologies only limited erosion has been achieved. When such forms are found stranded above present sea level as raised beaches they provide valuable evidence of fluctuations of sea level. In formerly glaciated areas, however, relative movements of the land with respect to sea level may result from either isostatic or eustatic causes. For the late Quaternary, eustatic changes are closely linked to the melting of glacial land ice, whose removal from the land causes isostatic rebound. In glaciated areas this can mean an immensely complex series of changes, especially as glaciers may have been debouching their outwash deposits into valleys flooded with the sea, so that outwash and beach deposits may be difficult to distinguish.

In Scotland these relationships have perplexed geomorphologists since the early nineteenth century. The finding of raised beaches with recent, but arctic, fauna appeared to imply higher sea levels in the late Quaternary. Only the acceptance of the isostatic model and the eustatic model together, and they were both proposed by Scotsmen, could resolve the problem. This acceptance did not really come about until Sissons began his campaign of precise levelling on Scottish raised beaches in the early 1960s. Conventional thinking had emphasized three levels of supposedly horizontal beaches at 100, 50 and 25 feet above sea level. However, levelling to Ordnance Survey benchmarks soon revealed that the beaches had strong tilts and that these reflected isostatic uplift since their formation. By tracing the shorelines along the coast and up the valleys the relation to the stages of glacial retreat could be established. It is now possible to see that the cold water fauna lived in beaches depressed both isostatically and eustatically below the present sea level.

The present pattern of beaches still presents problems. At several places there is unmistakable evidence of glacial deposits lying upon raised rock platforms and preserving beneath them beach sediments. Sissons (1967) thought that rock platforms close to present sea level were far more eroded than platforms due, undoubtedly, to postglacial seas. Consequently it looked as though the effect of the present wave environment, despite its obstensible power, was either minimal, or reflected its short stay at present levels. Still more puzzling was the width of the raised platforms, often in very sheltered localities in which it is difficult to imagine that the wave environment was ever greatly different to that at present. Sissons (1982) now believes that the higher raised beaches also need re-evaluation and re-mapping.

Sissons' work in Scotland shows that for some problems specific surveying techniques pay handsome dividends. Combined with reconnaissance mapping from air photographs, the exact delimitation of beaches is one example of the importance of exactitude in morphological mapping. The technique is so sensitive that recently it has revealed small postglacial faults displacing the famous pro-glacial lake shorelines of Glen Roy (Sissons and Cornish 1982), a presumed consequence of isostatic uplift following deglaciation.

The origin and migration of barrier islands

The coast of the United States is fringed with marginal linear islands termed barrier islands, and it is estimated that 13% of the world's coasts are fringed with barrier islands and their concomitant lagoons. An extensive literature has discussed their origin, and on a soft rock coastline their continued migration shoreward ensures that they are studied within the context of present processes.

The controversy over their origin has been maintained for decades. They were initially thought to have formed as offshore bars that were accreting to form subaerial ridges or sand barriers landward of the breakpoint of waves (de Beaumont 1845). Gilbert (1885) ascribed their origin to longshore drift and the elongation of spits in a shore-parallel direction. Barrier formation as wavelain features on an emerged continental shelf was proposed by Johnson (1919) but it proved impossible to show that shore processes alone could raise them above sea level during a phase of rapid sealevel rise. Another school of thought related them to a transgressive postglacial sea on a gradually shelving landsurface liberally supplied with suitable sand-sized sediments (Shepard 1960, Fisher 1968). Swift (1975) developed this further by pointing out the potential for reworking these sediments by episodic landward migration of a barrier island during transgression and vertical growth during stillstand. On the basis of deep borings and radio-carbon dates it seems as though the majority of them have formed in the last 4000 to 6000 years, i.e. since the sea reached about its present level, and they may now be migrating shoreward and wasting away by erosion of the seaward side and deposition in the lagoon behind them, unless the offshore or longshore supply of sand is sufficient to maintain them in steady state. Their ubiquity certainly calls for a common explanation and the public need for shoreline access ensures that the coastal zone will remain one of active controversy.

Landscape evolution over cyclic time

If, by common agreement, the Davisian model was judged an unsuitable framework for the explanation of regional landscapes over cyclic time, there has been far less agreement about what should replace it! I reviewed Hack's theory of dynamic equilibrium in the last chapter. Its advocacy of non-cyclic landforms, a type of perpetual maturity, was an intellectual challenge to Davisians and it was met by at least one rebuttal: Harlen Bretz's study of the Ozarks. Bretz (1962) wisely noted that 'neither the Davis sequence nor that espoused by Hack can be demonstrated from observations through time nor proved by continuously operating models or laboratory experiments.' This presages Bishop (1980) who accused the Davisian model of being unfalsifiable, although this was partly on the basis of internal, circular definitions of terms.

The resolution of some of the differences between process workers and cyclic thinkers was achieved by Schumm and Lichty's paper (1965), and Schumm as much as anybody has worked to keep cyclic thinking active. In 1963 he reviewed 'The Disparity between the Present Rates of Denudation and Orogeny' and showed that in active tectonic environments, erosion was at least an order of magnitude slower than uplift, and provided some comfort to Davisian laggards: the much-criticized simplifying assumption of rapid initial uplift followed by stability and erosion was correct to a first order approximation. He also graphed the effect on the Cycle of the isostatic response to long continued erosion.

Subsequent work by Schumm identified the notions of 'complex response' and 'thresholds' and attempted to replicate, by physical modelling, stream incision and basin response to changes of base level. This work amounts to a modern version of the complications possible for the basic Davisian cycle, with a greater recognition of the intricate balance between the factors of erosion and deposition, and the rapidity with which change can be effected.

The Sixth Binghamton Symposium was devoted to theories of landform development and one issue of importance was the growing necessity to see regional landscape evolution within the diastrophic framework supplied by plate tectonics. One set of external thresholds, in the sense of Schumm, ought to be a consequence of the vertical movements associated with plate motion, and they ought to have the merit of being sufficiently regional to impress landscapes over considerable areas. The symposium also reviewed the standard twentieth century theories of landscape evolution: Davis, Penck, King and some of their kinks, such as lateral planation (Crickmay 1975), and contrasted them with the dynamic thinking of Hack (1975) and the thought of G.K. Gilbert (Pyne 1975). While there were no startling insights at the symposium it revealed an awareness that regional landscape evolution remained as a problem to be addressed and resolved (Higgins 1975).

A more deliberate attempt to examine the problems attending models of landscape evolution was made by Brunsden and Thornes (1979). They attempted to determine whether there existed a set of general concepts that would allow the formulation of a rational basis for studies of long-term landform evolution. They examined the propositions that (1) constant processes will, over sufficient time, produce characteristic forms, that (2) perturbations to the system may or may not trigger transient behaviour in which the system shifts to another equilibrium state, that (3) 'successful' perturbations may trigger complex response, either as a chain of responses to a single input, or as spatially diverse results to the same input, and finally that (4) a landscape can be measured in terms of its sensitivity to change. They then suggested how these principles might be applied within a standard study of denudation chronology. The principles draw together existing conceptual thinking but no guarantee was offered that they form a sufficient set to guide all future thinking.

The paper appeared in a British journal sandwiched between a reassessment of the evolution of the shape of Britain by Brown (1979), and a consideration of the evolutionary geomorphology of Australia and Papua-New Guinea (Ollier 1979). Both writers place their studies within the framework of plate tectonics, which controls the general pattern of diastrophism in their respective areas. However from Brown's study it is clear that little topography older than 10 to 20 millions years old could have survived in Britain, and even from this period the primary evidence is tropical weathering products preserved in solution pockets. In contrast, Ollier described the evolution of one of the oldest continuous land surfaces in the world, some of which dates back to Triassic times, at least 200 million years, and predates the break-up of the super-continent Gondwanaland from which Australia emerged. Ollier rejected dynamic equilibrium, cyclic theories (except perhaps for the later Tertiary), and climatic geomorphology as bases for the study of Australian landforms. He regarded the surface as being too old, and too little affected by present processes, for standard theories to be of great use in explaining the topography, and he styled 'the sort of geomorphology we see in Australia..evolutionary geomorphology, which is part of the story of the evolving earth.' Thus the unique events of a particular tectonic history create a strongly idiographic scheme. In a similar idiographic vein Ford et al. (1981) have been able to date the erosional age of relief in the southern Rocky Mountains via geochronologic methods applied to fossil cave deposits now stranded high up in the local relative relief. The results indicate a remarkably rapid rate of erosion locally, but unfortunately the method is restricted to suitable lithological locations.

Systematic accounts of regional landscapes continue to appear. Sissons' account (1967) of the evolution of Scotland's scenery was the first since the last edition of Geikie (1901), and a recent volume has updated thinking on the evolution of southern England (Jones 1980). A comprehensive regional geomorphology of the United States (Thornbury 1965) bears witness to a continued interest, always strong in the United States, in regional physiography. However, such accounts are often difficult to reconcile with advanced thinking on systematic topics, and the problem is all the more acute during a period when the majority of work is directed toward medium to short timescales, and corresponding spatial scales. As a result significant lacunae still exist in connecting these two facets of geomorphology.

IN CONCLUSION

It is not possible to conclude this section in the usual sense since no obvious break point in geomorphological thought about cyclic time has yet occurred, at least not one than can be unambiguously detected at present. Bishop (1980) touched a sensitive point in noting that the Davis model cannot be falsified and is therefore unscientific. Bretz (1962) made essentially the same point about both Davis and Hack. If, in Hack's theory, equibilibrium processes cause the landscape to simulate Davisian maturity, how can we choose between alternative models? The postulate that the landscape 'always looks like that' is not greatly different from, say, a creationist argument that it was created to look like that: radiometric dates and all. It may be an argument from which there is no escape, and it remains to be shown that historical schemes in geomorphology are susceptible to objective testing. One possible solution to this dilemma involves the ergodic principle; that spatial change can be subsituted for temporal change. Savigear's paper (1952) is a famous example, but the principle can be used most securely over limited time spans and it still suffers from the problem that it is difficult to be sure what processes have actually acted over cyclic time, and whether they are statistically similar to those that vary over a limited spatial range. In a recent examination of the problem Craig (1982), while demonstrating that three-dimensional landforms evolving in time cannot be stable, concluded that the ergodic hypothesis 'allowed useful models of landform behaviour.'

More constructive still is the attempt by Hey (1979a) to model mathematically the development of landscapes with assumptions reflecting present knowledge. His results, for example, caused him to note that 'landforms produced during the initial phase of development, when local weathering was non-limiting, dominate the landscape' and he observed that this could cause the misfit condition for streams in the landscape without reference to climatic change. He also found the need (1979b) to question the causal assumptions in Schumm and Lichty (1965) and this may herald new lines of thought.

I intend to reserve the more general conclusions to be drawn from this survey of the history of geomorphology for the next and final chapter. At this point it suffices to observe that in the early 1980s geomorphology has apparently begun to achieve a balance between the amount of work devoted to process versus form, and to short versus long timescales. There can be no ideal agreeable to all parties, but there is little doubt that the merits of equilibrium in this matter, with perturbations coming from all sides, are appreciated by the majority of practitioners in the discipline.

Part V

ONE MORE CENTURY?: MAKING THE SYSTEMS WORK

The study of the complex interactions of all the competent processes tends to be neglected, though it is this interplay alone which leads to the creation of relief... Only the natural relief can be used for this and it cannot be replaced by the cleverest of simulations.

(J. Büdel, 1982)

Chapter Twelve

WHAT NEXT?

> *A system is thus formed, in generalising all those different*
> *effects, or in ascribing all those operations to a general end.*
> *This end, the subject of our understanding, is then to be*
> *considered as an object of design;....*
>
> (James Hutton, 1795, II-566)

Any attempt to see where present trends in geomorphology may lead us must base itself on a critical assessment of the tribulations that have beset the course of thought up to the present. What lessons can be learnt from this review of the discipline?

LESSONS FROM THE PAST

The Ancient World

The most striking fact about earth science in the Ancient World is the degree to which it failed to develop, despite the advantages of an active physical environment and the intellectual power of logical thought. In Chapter Two I suggested that this was due to a failure of the fundamental physics. Seneca and Strabo both stated the law of uniformity as explicit as Lyell, but unfortunately the laws that, for example, Seneca thought were continually working in an unchanging manner below the earth's surface were the wrong laws! Strabo's views were more acceptably modern, as far as we can tell, but with no experimental basis to test the physics there was no final arbiter as to whose system was correct. Physics in the Classical World had a habit, like Humpty Dumpty's words, of meaning exactly what an author wanted it to mean: an anarchy from which there was no escape.

An auxiliary point is that the Ancients do not appear to have developed the law of super-position, and there is little, if any, reference in Classical literature to the observation of different rock layers: the basis of a knowledge of stratigraphy. Without a general theory of geology the dynamics of the surface cannot be properly understood, and the Ancients were content to allow geological dynamics drive their physics; volcanoes and earthquakes, rather than the reverse. Thus we must conclude that the failure in the Ancient World was the failure to develop an acceptable discipline of physics and an experimental methodology. The lesson for the present may be that we must adopt the proper forms of physical models if we are to make substantial progress, a point that bears on the relation of geomorphology to civil engineering, discussed below.

Leonardo da Vinci to Hutton

The period of the Renaissance to the end of the eighteenth century was a period of intellectual recovery and one could take Aristotle's earth science and weld it onto late seventeenth century science in England during the early days of the Royal Society, and the seam scarcely would be visible. In both periods there coexisted the same easy mixture of common sense and global magic. On this view Hero's temple steam engine parallels the early steam engines in Britain, still a century away from general utility. It was to take at least a century from the death of Newton (1727) before the science of physics developed a unity that forced all scientists to take note of its findings and operate their systems within its framework. Only with this interpretation can one understand the debates between, for example, Hutton, De Luc and Whitehurst. It also reinforces the point just made about the physics of the Ancient World, and illustrates the magnitude of the step that they failed to make. In geology, this imperative was mainly felt through the slow realization that the field evidence was paramount, and that realistically this must be related to just those surficial processes that could be observed or reasonably supposed. The lesson is, therefore, to heed the over-riding importance of field evidence. A twentieth-century controversy that highlights exactly this point is the prolonged debate over the reality of the Spokane Flood and the origin of the Channeled Scabland (Clayton 1970a).

The nineteenth century

The fluvialism stated by Hutton and Playfair in the last decade of the eighteenth century was not fully adopted until the last few decades of the nineteenth. It is a mistake to see this century as a titanic struggle between good (fluvial uniformity) and evil (variegated catastrophism). As I have explained in earlier chapters, there were very good reasons for objecting to the fluvial doctrine in the early decades of the century. The wholesale and unthinking adoption of the idea would have been just as disastrous to progress as a further entrenchment of the various flood models then being proposed. The problem of erratic boulders, caricatures of the Drift, had to be disposed of, and the truly heroic battle of the century was fought over the Ice Age.

Lyell has often been castigated for his failure to

embrace the glacial theory, and his influential position and prominent writings certainly swayed the course of fluvialism and glacialism after 1830. Yet, his conservative position on both issues was well considered. After 1845 he was, at least to geomorphology, a follower rather than a leader, and his main interests lay elsewhere in geology. The field evidence for a glacial, or late glacial submergence did seem compelling in Britain, and both Andrew Ramsay and James Geikie believed in a period of submergence during their initial development of glacial stratigraphy. Lyell's support of fluvialism went as far as could be expected: the period from 1845 to 1860 was a muted one for fluvialism and it is significant that the next step forward was a broadly conceptual one; imagining the entire course of fluvial erosion on a landmass of varying geology. This notion, in turn, was tied intimately to the recognition of the importance of glacial erosion. Can we, should we, criticize Lyell in his cosmopolitan maturity for failing to have two crucial and brilliant ideas?

The nineteenth century demonstrates that the development of thought proceeds by faltering steps and that the real position is usually more complex than may be realized at the time. The principle of uniformity was a powerful antidote to magical catastrophism, but its real power only became evident when it was tied, in the imagination, to the working of processes over global space and geological time.

The twentieth century

Geomorphology in the twentieth century can scarcely be thought of separately from the name of W.M.Davis. The severely critical reaction to his influence was paramount after World War II and peaked in the early 1960's, since when a more moderate attitude towards him has prevailed thanks to the major biography provided by Chorley, Beckinsale and Dunn (1973). Chorley himself was a fierce critic of Davis so that the moderation of attitude is significant, and resulted in part from the task of reading Davis' prolific output.

In the rush to reject Davis during the second half of this century, many of the critics forgot that they also subscribed to the view that all that Davis had done was to synthesize the work of others. It was widely admitted that he had taken the works of Newberry, Powell, Gilbert, McGee and Dutton and used them to produce the Cycle. But, if we reject Davis, are we not forced to re-interpret the work of his predecessors? What then becomes of base level and grade? If we also look forward in time, how do the ideas we propose mesh with those of the past?

When a new paradigm is adopted in a scientific discipline the immediate urge is to reconstruct the history of the subject. Neglected sources are discovered, new heroes are adulated, and the discipline leapfrogs the immediate past in search of a more respectable ancestry. The emphasis on short term processes and systems theory that developed in the 1950s and 1960s found an ideal nineteenth-century secular saint in G.K.Gilbert. Yet, in leapfrogging Davis the modern period has overlooked, or undervalued, the important beginnings made on process studies in the 1930s, and the substantial alternatives provided by climatic geomorphology.

However, real progress is usually made by grafting new ideas onto old rootstock, and the new subsumes the old. A substantial amount of work is now being done on processes over graded time and this induces some consideration of the implications of this work for cyclic time. As a consequence it is clear that some decision now has to be made about the validity of the old models purporting to describe cyclic time. A serious re-evaluation of previous work should be undertaken and a reconcilation effected between the old and the new.

The lesson from the present century is that we should not reject too readily the work of the recent past, and that there should be a more conscious integration between it, and new trends in thinking. If we do not do it, historians of the subject in the future will do it for us.

Hidden and peripheral authors

It is not always the case that the most visible authors in an era are those who, potentially and from the point of view of posterity, ought have had the most influence. The Classical world is too remote for us to make any clear judgement, but it seems very likely that there may have been someone whose ideas were at least as sound as those of the early modern era, perhaps Pythagoras or Archimedes?.

In the Renaissance, Leonardo da Vinci, on the evidence of his own unpublished notebooks, might have founded land-scape science in 1500 much as it was around 1800, or even later. Robert Hooke, curator to the Royal Society, influenced later ages with his posthumous works, but he made no impact on the popular ideas of his contemporaries, Burnet, Woodward and Whiston. Similarly, Gautier's uniformitarian views lay at the roots of a French school of actualism in the eighteenth century, but they have only just been noticed (Ellenberger 1975).

The danger that the ideas of a radical thinker could be completely lost diminished as science became a more public enterprise, and capable of tolerating widely divergent views. Hutton is an example of a writer who might have been lost had he not been located in the intellectual ferment of the Scottish Enlightenment: Edinburgh in the late eighteenth century. Although his work was probably better known to his contemporaries than some writers have thought, his contribution, happily, was immortalized by his friend Playfair. In contrast, Lamarck, who published his 'Hydro-geologie' in the same year as Playfair's 'Illustrations' (1802) suffered almost total neglect, although his other work was well known. Carozzi remarks, in the introduction to his modern translation, that some of Lamarck's biographers have even doubted the existence of the small, privately published volume!

On the basis of modern judgement G.K.Gilbert might be identified as an author, well enough known on the one hand, but curiously neglected on the other. He was highly regarded by W.M.Davis, who, however, could make little of Gilbert's neglect of the time factor in the analysis of landscape; the very trait that was to appeal to the modern age. Gilbert spent his entire career with the U.S.Geological Survey and wrote no books so that his ideas had little chance to gain a group of dedicated disciples. In the twentieth century Walther Penck comes into neglected category, at least from the point of view of his homeland. Bremer (1983) has documented the neglect of W. Penck in Germany right down to the present time, despite the considerable attention paid to his theories in the English speaking world. This fact provides an interesting reflection of the intellectual polarization existing amongst national schools of geomorphology.

It is at least worth a thought as to who might be a parallel figure today. The traits that the under-rated or neglected author displays are not easily recognized, not least for that very reason; the central relevance of what they are saying is not evident, or is difficult to assimilate into contemporary thought. However, on the basis of the names just reviewed it may be noted that such an author is often well-known and/or well-placed, yet slightly peripheral to those who would identify themselves as the core of the progressive establishment: da Vinci, Hooke, Gautier, Hutton, Lamarck, Gilbert and Penck. My own nominations for the modern period would be R.A.Bagnold and W.E.H.Culling. The future may find that their methodology is as critical to future progress as their actual results.

Geomorphology and diastrophism

The delicate balance that exists in the landscape between uplift and downwearing has been a persistent theme underlying geomorphological thought. In any consideration of the evolution of landscape over anything more than a few years, the prevailing form of diastrophism, either as a conceptual model or as actual motions of the earth's crust, makes itself felt. In consequence it permeates both the theory and the practice of geomorphology. The nineteenth century in particular, whose first sixty years was a period of

diastrophic anarchy, eventually was forced to come to grips
with accumulating geomorphological evidence which revealed
slow crustal movement, such as isostatic rebound in
Scandinavia, and to incorporate it into an acceptable theory
of crustal mobility. The failure of nineteenth-century
geology, relatively speaking, to develop a diastrophic model
close to what we would now regard as the 'truth' naturally
hampered the proper development of geomorphology during that
century. The diastrophic assumptions of Davis, although
simplistic, were entirely in accord (p117) with existing
theory, and yet were scarcely unchanged from those implied by
Hutton the century before. Late nineteenth-century
diastrophism was so rudimentary that Hutton's rock-cycle, it
may be argued, fits more harmoniously with the prevailing
paradigm of plate tectonics than with any previous scheme.
Davis assumed simplicity, although both he and Penck
subsequently imagined more complex scenarios. Even now the
early stages of mountain building and subaerial uplift are
shrouded with mystery, at least at the time-scales needed for
geomorphic modelling.

There can be no doubt that the lesson to be learnt is
that geomorphology ignores diastrophic theory at its peril,
however tenuous the connection might appear to be on the
surface. From this viewpoint it is interesting to note that
Budel (1982) ignores plate tectonics in developing climatic
geomorphology, referring instead to 'random diastrophism.'
Likewise the Russians have eschewed plate tectonics for the
empiricism of neotectonics, and a continued belief in the
efficacy of mechanisms which produce widespread planation.

Geomorphology and civil engineering

A recurrent theme from the earliest times has been the link
between the practise of civil engineering and geomorphology.
In the Ancient World it is likely that civil and military
engineers were the practicing geomorphologists of the era but
the rules of thumb in daily use are as easily lost as
manuscripts, and the practical man is not given to writing
philosophical treatises on the implications of his diurnal
labours. Thus, our main knowledge of Ancient thought on
processes acting over cyclic time is derived mainly from
philosophers who sneered at manual mechanics.

Leonardo da Vinci headed a long line of Italian
engineers up until the eighteenth century when the innovating
influences spread northwards to France and Switzerland. Earth
science, in general, ignored all that the engineers had to
say, and the dichotomy might be cast into a modern framework
by seeing it as a conflict in time scales: the engineers were
working with steady time, the global theorists with cyclic
time, or rather as they saw it, no less than the age of the
Universe. Gautier, the French civil engineer in charge of

roads and bridges was an interesting exception since he used his professional experience to write a formal theory of the earth based on a knowledge of diurnal processes, which he related also to their implications over lengthy time spans. However, as I indicated above, his work was neglected.

The impact of the civil engineers might have been profound at about the beginning of the nineteenth century if Playfair had managed to write the second edition of his 'Illustrations.' His 'Outlines of Natural Philosophy' (1814) was written for his Edinburgh students and from the incidental references to rivers and valleys in them it is clear that his revised 'Illustrations' might have gone a long way towards placing geomorphology on a sound footing. His awareness that not all valleys had been eroded by rivers, and his enlarged understanding of European glaciation (by 1816), suggest that his book would have been a model of clarity and perception. But he died before it was started and only the outline survives (p60).

As a result, engineering ideas had to begin anew the process of filtering into geological issues. Lyell referred to Robison's article on 'Rivers' in the 'Encyclopaedia Brittanica' which drew on the continental work, primarily Du Buat, but the material was used rather as useful evidence than as the basis of a methodology. Geology itself was concerned with much greater issues, in establishing the stratigraphy of the Earth, and fluvial doctrines were merely incidental issues to the great majority of geologists.

However, by the end of the nineteenth century the connections were once more established, and particularly in the United States where G.K.Gilbert epitomized the practical engineer's approach to a variety of geological problems including, for example, the structural problems posed by the laccoliths of his Henry Mountains report. The advent of W.M.Davis deflected the impact that this thinking might have had on geomorphology by reorienting the subject to the cyclic timescale once more. The dichotomy of this approach is well illustrated by Russell's book on rivers in North America (1898). The early portions might act as a nineteenth century 'Fluvial Processes in Geomorphology' (Leopold et al. 1964), but the author's real feelings are revealed in a positively melodramatic presentation of the Cycle in the last chapter.

Geomorphology and civil engineering parted company after 1900 and at least some of the slow progress of geomorphology up to 1945 has been blamed on this separation, and the supposedly baneful influence of Davis. However, the first half of the century was beset by two World Wars and the Depression decade of the 1930s. The actual number of professional geomorphologists was very small, and their teaching activities encouraged an attention to spans of cyclic rather than to steady time. The move to return to the basic mechanisms of process had already begun before World

War II (p161), and it is arguable that geomorphology was about to expand significantly at the end of the 1930s. A 'Journal of Geomorphology' began in 1938, several textbooks appeared from the new generation of geomorphologists (Wooldridge and Morgan 1937, Lobeck 1939, Von Engeln 1942, Cotton 1941, 1942, 1944, and King 1942) and it is a matter of speculation in what direction the subject would have progressed. At the very least, World War II probably caused the momentum of a decade to be lost and, with the general shifts in interest that resulted from the War, the real hiatus was considerably longer; perhaps two decades. If we make an allowance for this hiatus we arrive at the late 1950s by which time the subject had renewed its links with engineering via Horton's 1945 paper (p182) and the group at the U.S. Geological Survey.

The present day sees a curious ambivalence. Civil engineering might be regarded as the science (applied physics), of which geomorphology is the art. The modern urge for social relevance, encouraged by the tax-paying public, engenders attempts to retreat into environmental engineering as part of the price of respectability. Geomorphologists in the United Kingdom (Brunsden, Doornkamp and Jones 1978) have discussed the desirability of professionalizing the subject in line with other fields of science (other than geology). At the same time, increasing attention is being given to field problems embedded in graded or cyclic timescales, a topic upon which engineering has traditionally had little to say. The present institutional structure of science, powered by public funds, makes it unlikely that geomorphology will again lose touch with the engineers. However, if it turns out to be true that cyclic landscape behaviour is largely independent of daily processes, as Church (1980) suggests, then a total subjugation of the discipline to engineering principles would be a serious, and unnecessary, over-reaction.

THE PRESENT DAY: A PATHWAY TO THE FUTURE

The institutional position

It is difficult to give a synoptic view of modern geomorphology, but a number of characteristics can be pointed out. The last two decades have seen a marked expansion of the university system over most of the world, but the present recession is now eroding that position and a stabilization of, or even a slight reduction in, the number of practising geomorphologists in universities can be envisaged. However, the hidden benefit in this necessary evil is a breathing space for evaluation for those in the university system, and outside it a movement of trained geomorphologists into other related employment, particularly private and public agencies concerned with various aspects of environmental management. One might say that by the end of the eighteenth century a few individuals had trained themselves in landscape science; a century later a profession had trained itself. In the last decades of this century strenuous efforts are been made to ensure that the general public is educated about the workings of the earth's surface systems. The urge to educate the public is not a new one, it was common in Victorian Britain, but the realities of the practical management of landscapes frequented by large sections of the public are relatively new, and geomorphologists have a valuable part to play in publicizing the critical issues: for example, floodplain management and slope stabilization.

I have mentioned in the last chapter, and in the last section, some of the institutional changes that have characterized the last two decades. Geomorphology once more has its own English language journal, its own societies, and the first international conference in the subject is scheduled to be in Manchester (U.K.) during September 1985. Previously geomorphologists have had to meet under the umbrellas of geographers and geologists, with a consequent reduction and dilution in attendance. Viewed from the English speaking world, the catalyzing ideas for the discipline primarily appear to be those that are published in the major, English language, international journals. Despite the existence of the trilingual 'Zeitschrift fur Geomorphologie,' and several excellent French journals, e.g. 'Revue de Geomorphologie Dynamique,' the interchange of ideas between languages still is regrettably limited (Clayton 1970b). It is difficult to say whether this represents truly the balance of seminal ideas in the geomorphological literature; it is to be hoped that it does not. Certainly, the average paper in geomorphology cites few, if any foreign language papers, although the availablity of Geo Abstracts leaves no real excuse for ignorance. As a random example, the Binghamton Symposium volume edited by Thorn (1982) has sixteen papers citing well over three hundred references. Of these, only four are foreign language papers, and three of these are cited by one paper. Such an imbalance may merely reflect the use of English as the international language of science, but it is certainly food for thought. Likewise only about 16% of Budel's (1982) bibliography is non-German language literature and large chunks of English language material relevant to his themes are omitted, e.g. the work of Dury on misfit streams. The personal contacts of international conferences should provide the lines of communication that may be lacking at present, and the formal presentations should provide the basis, at least in principle, for a global unification of geomorphological thought. Perhaps one benefit to be gained soon from the current electronic revolution might be the ready availability of machine translations of scientific papers. Despite the sterling efforts of Geo Abstracts, it would appear that a translated abstract is insufficient to ensure the diffusion of new knowledge when the source is in another language. This can be readily understood when it is appreciated that the skeleton of an idea is usually insufficient in itself: an acceptance or refutation requires a considered view of all the supporting evidence. Inertia and environmental bias are usually sufficient to prevent the pursuit of a full translation.

The beginning of the nineteenth century saw great structural changes in the organization of science. These accompanied, or even preceded, revolutionary changes in the conduct and methodology of most of the sciences, including geology. Geomorphology slowly emerged as an independent entity by the end of that century. Publicly funded science has restructured itself in recent decades, both nationally and internationally, and geomorphology is once more emerging a little behind geology but as a recognizable subject. These structural changes do seem to have been accompanied, as in the last century, by a spurt in the growth of the discipline, but it is too early to assess this growth for its real significance.

The data base and future conceptual progress

One feature of contemporary science is the availability of increasingly detailed data sets, and extremely sophisticated computing facilities. The automatic monitoring of diurnal processes in real time to a host computer, whether from slope, river or satellite, is becoming a serious possibility and is both convenient and expensive. It brings with it the threat of 'information overload' but properly handled, it ought to provide the continuous observational basis that the subject, despite strenuous efforts, has lacked until the present time. Welded to simulation techniques, high speed computing, and visual displays there exists the theoretical potential to make landscapes work, mathematically, under any chosen set of environmental controls. Rudimentary models of this type have already been developed and it is conceivable that the next step forward may be the formal operation of landscape systems in numerical practice, illustrating and subsuming theoretical principles developed over two centuries. An essentially numerical approach may even be demanded by the subject itself since it appears that analytical solutions to the characteristic form problem of slopes in three dimensions are intractable (Thornes 1978).

The reasoning behind this conception of the future is as follows. In the broadest possible terms major changes in the controlling paradigms of geomorphology have been made by asserting very generalized geometries; i.e. by establishing a controlling topology for the system.

In the first instance this was extremely simple, but not without power. Hutton maintained that after uplift, gradual downwearing by subaerial action reduced a land surface to a plain. Clearly, this implies virtually nothing about the precise geometric forms that will be encountered. Objections to this scheme followed two general lines: one asserted that a surface, possibly after a 'settling in period,' maintained a static state undergoing no morphogenesis. The other proceeded by finding geometries which the Huttonian model could not predict: lake basins, of which the most telling examples were glacially excavated rock basins like Lake Geneva. The operation of surface processes in the Huttonian theory ought to produce, logically, a continuously sloping surface from an arbitrary initial surface. Hutton and Playfair both sought to overcome the problem of Lake Geneva by expanding the topological basis in an ad hoc manner; by postulating tectonic or karstic agencies for example. Hutton, just as much asPlayfair, understood the seriousness of this particular objection.

Further progress was only possible by removing this objection, which was eventually done by developing a glacial theory free from these constraints. The glacial theory permitted an even more general geometry to be added to Hutton's topology, since closed hollows and rock basins below sea level were no longer a problem. The problem, indeed, was defining the geometric limits of this new freedom.

Not long afterwards Davis reduced the field to order by generalizing the geometry again with the addition of the temporal dimension, now specified whereas previously it was left unfettered. He further clarified matters by expanding the climatic scope of the topology and by permitting the system to repeat or renew itself in response to arbitrary exogenous inputs. Some of these interruptions re-started the Cycle, others produced different geometries: glaciation, extreme aridity, and volcanism. For the important fluvial system he specified a generalized and deductive space/time geometry: universal downwearing with certain geometric aspects of slope and river profiles specified in terms of convexity and concavity.

As with Hutton, so with Davis, objectors either denied the existence or importance of the temporal dimension of the topology (for the fluvial model and its derivatives) and produced the steady state landscape (Ashley, Fenneman and Hack), or they objected to specifics of the space/time geometry (Penck).

The modern period is still grappling with the resolution of this problem but already attempts have been made to generalize the space/time geometry by showing how some of the observed discontinuities, partial cycles, can be generated by isostatic unloading, or by graded time imbalances between erosion, transportation, and sedimentation. While Davis merely stated that a landform system existed, this century has been concerned to discover the interconnections between different parts of the system, and we are slowly arriving at a point at which quantitatively based statements can be made about these connections.

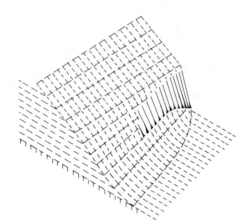

Figure 17. *In current work M.J. Kirkby uses a numerical computer model to simulate ten thousand years of slope development on part of the cliff profiles in South Wales originally studied by Savigear (1952). The right-hand side of the diagram has undergone ten thousand years of simulated cliff erosion. The left-hand side underwent five thousand years of cliff erosion followed by five thousand years of slope erosion at realistic rates for soil creep and landslips, the two main processes recorded in the area. (Scale ticks and contours at 5 metre intervals).*

If we are to carry the analogy further then we should use all the facilities available to us to exploit the topology of the system that we have just begun to establish, now a step removed from the specific geometries of landscapes. It is at this point that we return to the data base, and the high speed computer with high resolution graphics! The exploration of the system phase space, under the myriad controlling inputs that are possible, ought to reveal any discontinuities that exist as a function of the controls. A similar approach, in principle, should also be possible for coastal and glacial geomorphology. Karst geomorphology may turn out to be a by-product of the fluvial model for a particular set of controls. The idea that a paradigm for geomorphology is merely an evolving image on a screen which replicates reality may be offensive to the field geomorphologist (pace Budel), but it can hardly offend us more than Hutton's simple notion that the earth wore down offended his contemporary zealots. Davis' scheme, after all, amounted to no more than imaginary snapshots of just such a moving film, and the late twentieth century is certainly inured to accepting the dominating influence of the small screen. If the idea seems overly simple we can reflect that the 'simple' ideas of Hutton and Davis have each powered a century of research and controversy.

However, merely getting a landscape to evolve on a screen will not complete our researches, it is only a theoretical prescription. The system structure is still far from specified in many of its details and the field itself still has much to teach us. There is no need for the discipline to feel defensive about the fact that it may be about to enter a renewed period of theorizing. If it is about to make this move, and if it does, it will be merely symptomatic of a condition in which the subject finds itself. No discipline can afford to neglect useful tools from feelings of simple prejudice, and internal maturity ought to be sufficient to keep theory and data in a proper proportion.

Field work on other planetary bodies, sometime in the next century, will extend the scope of the discipline, and speculation is already underway about the controlling forces affecting martian landforms. The controlled conditions of radically different environments will inevitably teach us something about our own planet, just as earlier explorations on our planet have done, and underlying everything will be the spatio-temporal dynamics from the prevailing diastrophic models of geology.

TO CONCLUDE

Geomorphology is, and always has been, the most accessible earth science to the ordinary person: we see scenery as we sit, walk, ride or fly. It is a part of our daily visual imagery, and we do not even have to stop or stoop to examine it, although our perceptions are usually better if we do. Extraordinary passions have been aroused over the centuries by mundane objects, such as large rocks containing minerals foreign to a locality. We must hope that such problems and passions will continue to arise. Nature has a way of teaching us what she will: beauty, truth and knowledge all lie in the eye of the beholder and provided that we return to the field we shall not languish from a shortage of inspiration:

> I do wonder and puzzle whether there is a divine design at the heart of the beauty of slope and shore.

This might have been James Hutton, but it was written in 1979. I leave it unacknowledged as an unconscious tribute from our age to a founder of our discipline: the teleological urge may tell us nothing, but it has the power to motivate beyond the call of duty.

BIBLIOGRAPHY

Abbreviations

AAAG	Annals Ass. Amer. Geog.	IBGT	(ditto) Transactions
AGS	Amer. Geog. Society	JG	Journal of Geology
AJS	Amer. Journal of Science	nf	new folio
BEB	Beach Erosion Board	NGM	National Geog. Mag.
EMG	East Midland Geographer	NGS	National Geog. Society
ENPJ	Edinburgh New Phil. journ.	ns	new series
ESP	Earth Surface Processes	PG	Professional Geographer
ESPL	(ditto) & Landforms	PGA	Proc. Geol. Assoc.
GJ	Geographical Journal	QJGS	Quart. J. Geol. Soc.
GM	Geological Magazine	QR	Quaternary Research
GR	Geographical Review	RSET	Roy. Soc. Edin. Trans.
GSA	Geol. Society of America	SGM	Scottish Geog. Mag.
GSAB	(ditto) Bulletin	USGS	U. S. Geol. Survey
IBG	Inst. Brit. Geographers	USGSPP	(ditto) Prof. Paper
	ZfG Zeitschrift fur Geomorphologie		

ADIE, A.J., 1835, On the expansion of different kinds of
    Stone from an increase in Temperature, with a Description
    of the Pyrometer used in making the experiments, RSET,
    13, 354-372
AGASSIZ, L.J.R., 1840a, On the polished and striated surfaces
    of the rocks which form the beds of the glaciers in the
    Alps. Proc. Geol. Soc., 3, 321-2
---, 1840b, Etudes sur les Glaciers, Neuchatel (rprt 1966
    Dawson:London, also Eng.trs. CAROZZI, A.V., 1967,
    Hafner:New York)
---, 1847, Systeme glaciaire, Masson:Paris
AGRICOLA, G., 1546, De Ortu et Causis Subterraneorum, Basle
ALBRITTON, C.C. (ed.), 1963, The Fabric of Geology,
    Addison-Wesley:Reading (Mass)
---, (ed.), 1975, Philosophy of Geohistory, Benchmark papers
    in Geology 13, Dowden, Hutchinson & Ross
    Inc.:Stroudsburgh (Penn.)
ALEXANDER, D., 1982, Leonardo da Vinci and fluvial
    geomorphology, AJS, 282, 735-755
ALFORD, J.J., 1982, San Vincente Arroyo, AAAG, 72, 398-403
ALLAN, A., 1815, An account of the mineralogy of the Faroe
    Islands, RSET, 7, 229-267
ALLISON, I.S., 1933, New Version of the Spokane Flood, GSAB,
    44, 675-722
ANDERSEN, S.A., 1931, The waning of the last continental
    glacier in Denmark as illustrated by varved clays and
    eskers, JG, 39, 609-624

ANDERSSON, J.G., 1906, Solifluction, a component of subaerial
    denudation, JG, 14, 91-112
ANDREWS, J.T., 1970, Post-glacial uplift in Arctic Canada,
    IBG Spec. Pub. 2
---, 1975, Glacial systems, Duxbury:Boston
ANHERT, F., 1976, Brief description of a comprehensive three
    dimensional process-response model of landform
    development, ZfG, 25nf (supp.), 29-49
ANON (A.B.), 1768, The wonders of Canada. A letter from a
    gentleman to the Antigua Gazette, New York Aug 21, 1768
    (rprt Mag. Am. Hist., 1, 243-246)
ANSTEAD, D.T., On the phenomena of the weathering of rocks,
    illustrating the nature and extent of sub-aerial
    weathering, Trans. Camb. Phil. Soc., 2, pt.2, 387-95
ANTEVS, E., 1922, Recession of the Land Ice in New England,
    GSAB, 33, 86 (Abstract)
ARBER, M.A., 1940, Outline of south-west England in relation
    to wave attack, Nature, 146, 27
---, 1949, Cliff profiles in Devon and Cornwall, GJ, 114,
    191-7
ARDEN-CLOSE, C.F., 1926, The early years of the Ordnance
    Survey, (rprt 1969 with intro. by HARLEY, J.B. and index,
    David & Charles:Newton Abbott)
ARISTOTLE, (1952), Meteorologica, Eng. trs. and intro. by
    LEE, H.D.P., Harvard UP:Cambridge (Mass)
ASHLEY, G.H., 1931, Our youthful scenery, GSAB, 42, 537-45
---, 1935, Studies in Appalachian mountain sculpture, GSAB,
    46, 1395-1436
ATWOOD, W.W., 1940, The physiographic provinces of North
    America, Ginn and Co:Boston

BAGNOLD, R.A., 1933, A further journey through the Libyan
    Desert, GJ, 82, 103-29
---, 1935, The movement of desert sand, GJ, 85, 342-369
---, 1937, The transport of sand by wind, GJ, 89, 409-438
---, 1940, Beach formation by waves- some model experiments
    in a wave tank, J. Inst. Civil Eng., 15, 27-52
---, 1941, The physics of blown sand and desert dunes,
    Methuen:London
---, 1946, Motion of waves in shallow water, Proc. Roy. Soc.
    A, 187, 1-15
---, 1947, Sand movement by waves; some small-scale
    experiments with sand of very low density. J. Inst. Civil
    Eng., 27, 447-469
---, 1960, Some aspects of the shape of river meanders,
    USGSPP, 282-E
---, 1966, An approach to the sediment transport problem from
    general physics, USGSPP, 422-I
BAILEY, E.B., 1967, James Hutton - The founder of modern
    geology, Elsevier:Amsterdam

BAKER, V.R., 1973, Erosional form and processes for the catastrophic Pleistocene Missoula floods in eastern Washington, in MORISAWA, M. (ed.) 1973, Fluvial Geomorphology, State Univ. NY:Binghamton, 123-148

---, 1978a, The Spokane Flood Controversy and the Martian Outflow Channels, Science, 202, 1249-1256

---, 1978b, Adjustment of fluvial systems to climate and source terrain in tropical and subtropical environments, in 'Fluvial sedimentology', Canadian Soc. of Petr. Geol. Memoir 5, 211-230

BAKER, V.R. & KOCHEL, R.C., 1982, Palaeoflood hydrology, Science, 215, 353-360

BAKER, V.R. & NUMMENDAL, D., 1978, The Channeled Scabland, NASA:Washington

BAKER, V.R. & PYNE, S., 1978, G.K.Gilbert and modern geomorphology, AJS, 278, 97-123

BAKEWELL, R.(sr), 1813, An Introduction to Geology, (later eds: 1815, 1828, 1833, 1838, American eds 1829, 1834, & 1839 include appendix by SILLIMAN, B., Rprt 1978 of 1834 ed., Arno Press:New York)

BAKEWELL, R.(jr), 1830, On the Falls of Niagara, and on the physical structure of the adjacent country, Loudon's Magazine of Natural History, 3, 117-130

BARNES, F.A. & KING, C.A.M., 1951, A preliminary survey at Gibraltar Point, Lincolnshire, Bird Obs. and Field Res. Station, Gibraltar Pt., Lincs. Report for 1951, 41-59

---, 1961, Salt marsh development at Gibraltar Point, Lincolnshire, EMG, 2(15), 20-31

BARRELL, J., 1919, The status of the theory of isostasy, AJS, 48, 291-338

BARTLETT, W.H.C., 1832, Experiments on the expansion of building stone by variations of temperature, AJS, 23, 136-140

BASSET, C.A., 1815, L'explication de Playfair sur la Theorie de la Terre par Hutton, traduite de l'Anglaise et accompagnee des Notes, Paris

BAULIG, H., 1926, Sur une methode altimetrique d'analyse morphologique a la Bretagne Peninsulaire, Bull. Ass. Geog. France, 10, 7-9

---, 1928, Le Plateau Central de la France et sa bordure Mediterraneene, Armand Colin:Paris

---, 1935, The changing sea level, George Philips and Son Ltd:London (rprt 1956)

---, 1939, Deux methode d'analyse morphologique appliquees a la haute Belgique, Bull. Soc. Belg. Etudes Geog., 9, 165-84

---, 1959, Morphometrie, Ann. de Geog., 68, 385-408

BECKETT, P.H.T. & WEBSTER, R., 1965, A classification system for terrain, Rept. 872, MEXE:Christchurch

BECKINSALE, R.P., 1976, The international influence of William Morris Davis, GR, 66, 448-466

BEEDE, J.W., 1911, The cycle of subterranean drainage as
    illustrated in the Bloomingdale, Indiana, quadrangle,
    Indiana Acad. Sci., Proc., 20, 81-111
BEHRE, C.H., 1933, Talus behavior above timber in the Rocky
    Mountains, JG, 41, 622-35
BELOUSSOV, V.V., 1970, Against the hypothesis of ocean floor
    spreading, Tectonophysics, 9, 489-511
BENNETT, H.H., Soil Conservation, McGraw Hill:New York (rprt
    1970 Arno Press:New York)
BENTLEY, R., 1693, A sermon (being the fourth of the first
    Series of Boyle lectures, entitled 'A confutation of
    Atheism), London
BERNHARDI, R., 1832, An hypothesis of extensive glaciation in
    prehistoric times. Jahr. fur Min., Geognosie und
    Petrefaktenkunde, 111, 257-267
BINTLIFF, J.L., 1976, The plain of western Macedonia and the
    Neolithic site of Nea Nikomedeia, Proc. Prehist. Soc.,
    42, 241-62
BIRD, J.B., 1959, Recent contributions to the physiography of
    arctic Canada, ZfG, nf3, 151-74
---, 1967, The physiography of arctic Canada, The Johns
    Hopkins Press:Baltimore
BIROT, P., 1960, Le cycle d'erosion sous les differents
    climats, Paris (Eng. trs., JACKSON, I.J., and CLAYTON,
    K.M., 1968, Batsford:London)
BISHOP, B.C., Shear moraines in the Thule area, north-west
    Greenland, U.S. Snow, Ice and Permafrost Research
    Establishment Res. Rep. 17
BISHOP, P., 1980, Popper's Principle of Falsifiability and
    the irrefutability of the Davisian Cycle, PG, 32, 310-15
BLACKWELDER, E., 1933, The insolation hypothesis of rock
    weathering, AJS, 26, 97-113
BLOOM, A., 1963, Late-Pleistocene fluctuations of sealevel
    and postglacial crustal rebound in coastal Maine, AJS,
    261, 862-879
BONNEY, T.G., 1893, The story of our planet, Cassell &
    Co.:London
---, 1912, The work of rain and rivers, At the University
    Press:Cambridge
BOUGUER, F., 1749, La figure de la terre determinee par les
    observations de M. Bouguer et de la Condamine, Academie
    des Sciences:Paris
BOULTON, G.S., 1972, The role of thermal regime in glacial
    sedimentation, IBG Spec. Pap. 4, 1-20
---, 1974, Processes and patterns of glacial erosion, in
    COATES, D.R.,(ed.), Glacial Geomorphology, State Univ.
    New York: Binghamton, 41-87
BOULTON, G.S., JONES, A.G., CLAYTON, K.M. & KENNING, M.J.,
    1977, A British Ice Sheet model and patterns of glacial
    erosion and deposition in Britain, in SHOTTON, F.W.
    (ed.), British Quaternary Studies, Clarendon Press:

Oxford, 231-246

BOURNE, W., 1578, A Booke Called the Treasure for Traveilers, devided into five Bookes, or Partes, London

BOWMAN, I., 1926, The analysis of landforms; W. Penck on the topographic cycle, GR, 16, 122-32

BRADLEY, W.C., 1958, Submarine abrasion and wave-cut platforms, GSAB, 69, 967-974

BREMER, H., 1983, Albrecht Penck (1858-1945) and Walther Penck (1888-1923), two German geomorphologists, ZfG, 27nf, 129-138

BRETZ, J.H., 1923, The channeled Scabland of the Columbian Plateau, JG, 31, 617-649

---, 1925, The Spokane flood beyond the Channeled Scabland, JG, 33, 97-115

---, 1928, The Channeled Scablands of eastern Washington, GR, 18, 446-477

---, 1932, The Grand Coulee, AGS Spec. Pub. 15, New York

---, 1942, Vadose and phraetic features of limestone caverns, JG, 50, 675-811

---, 1953, Genetic relations of caves to peneplains and big springs in the Ozarks, AJS, 251, 1-24

---, 1962, Dynamic equilibrium and the Ozark land forms, AJS, 260, 427-438

BRETZ, J.H., SMITH, H.T.U. & NEFF, G.E., 1956, Channeled Scabland of Washington; new data and interpretations, GSAB, 67, 957-1049

BROWN, E.H., 1960a, The relief and drainage of Wales, University of Wales Press:Cardiff

---, 1960b, The building of southern Britain, ZfG, 4nf, 264-274

---, 1961, Britain and Appalachia: a study in the correlation and dating of planation surfaces, IBGT, 29, 49-66

---, 1979, The shape of Britain, IBGT, 4ns, 449-46

BROWN, R.J.E., 1960, The distribution of permafrost and its relation to air temperature in Canada and the USSR, Arctic, 13, 163-77

---, 1967, Permafrost in Canada, Canada, Geological Survey Map, 1246A

---, 1970, Permafrost in Canada, University of Toronto Press:Toronto

BRUNSDEN, D., DOORNKAMP, J.C. & JONES, D.K.C., 1978, Applied geomorphology; a British view, in EMBLETON, C. et al., 1978, 251-262

BRUNSDEN, D. & THORNES, J.B., 1979, Landscape sensitivity and change, IBGT, 4ns, 463-484

BRUUN, P., 1954, Coast erosion and the development of beach profiles, BEB Tech. Mem. 44

BRYAN, K., 1922, Erosion and sedimentation in the Papagayo County, Arizona, USGS Bull. 730B, 19-90

---, 1934, Geomorphic processes at high altitudes, GR, 24, 655-656

---, 1940, The retreat of slopes, AAAG, 30, 254-268

---, 1946, Cryo-pedology- the study of frozen ground and
    intensive frost action with suggestions on nomenclature,
    AJS, 244, 622-42

BUCH, L. von, 1824, Physicakalische Beschreibung der
    Canarischen Inseln, Berlin

BUCKLAND, W., 1820, Vindiciae Geologicae; of the Connexion of
    Geology with Religion, Oxford (read 1819)

---, 1823, Reliquiae Diluvianae, London

BUDEL, J., 1948, Das System der Climatischen Geomorphologie,
    Verhandl. Deutscher Geog., 27, 65-100 (Eng. trs. in
    DERBYSHIRE, E. (ed.), 1973, 104-130)

---, 1957, Die "Doppelten Einebnungsflachen" in den feuchten
    Tropen, ZfG, 1nf, 201-228

---, 1963, Klima-genetische Geomorphologie, Geographische
    Rundschau, 15, 269-85

---, 1982, Climatic Geomorphology, Princeton University
    Press:Princeton New Jersey

BUFFON, G.L.L. Count de., 1751, Histoire Naturale, Paris,
    (Eng. trs. by KENDRICK, W., 1755, Bell:London, Eng. trs.
    of 2nd ed. by SMELLIE, W., 1785, London)

---, 1778, Epoques de la Nature, Paris

BUNTING, B.T., 1961, The role of seepage moisture in soil
    formation, soil development and stream initiation, AJS,
    259, 503-18

BURNET, T., 1681, Telluris Theoria Sacra, W. Kettilby:London,
    2nd ed 1689, (Eng. trs. The Theory of the Earth, 1684 &
    1690)

BUTZER, K.W., 1958, Das okologische Problem der neolitischen
    Felsbilder der ostlichen Sahara, Abhl. Akad. Wiss. Litt.
    (Mainz) Math.-Naturw. Kl., 1, 20-49

CAIRNES, D.D., 1912, Differential erosion and equiplanation
    in portions of Yukon and Alaska, GSAB, 23, 333-348

CARLSON, C.A., 1950, Trends in geomorphic research, GSAB, 61,
    1169-70

CARLSON, C.W., 1969, Downstream variations in the hydraulic
    geometry of streams: special emphasis on mean velocity,
    AJS, 267, 499-509

CARPENTER, N., 1625, Geography Delineated forth in two
    bookes, Oxford

CARR, A.P., 1965, Shingle spit and river mouth: short term
    dynamics, IBGT, 36, 117-129

CARRUTHERS, R.G., 1953, Glacial drifts and the undermelt
    theory, Hill & Son:Newcastle upon Tyne

CARSON, M.A., 1971, The mechanics of erosion, Pion:London

CARSON, M.A. & KIRKBY, M.J., 1972, Hillsope form and process,
    Cambridge University Press:Cambridge

CARTER, J., 1959, Mangrove succession and coastal change in
    south-west Malaya, IBGT, 26, 79-88

CATCOTT, A., 1761, Treatise on the Deluge, M. Withers:London

CAXTON, W., 1481, The Myrrour of the Worlde, London

CHAMBERLIN, T.C., 1909, Diastrophism as the ultimate basis of
    correlation, JG, 17, 685-693

CHAMBERLIN, T.C. & SALISBURY, R.D., 1909, Geology (2nd ed.),
    John Murray:London (1st ed. 1904)

CHAMBERS, R., 1848, Ancient Sea Margins, W. & R.
    Chambers:Edinburgh

CHAPMAN, L.J. & PUTNAM, D.F., 1951, The Physiography of
    Southern Ontario, Ontario Res. Foundation at Univ.
    Toronto Press:Toronto (2nd ed. 1966)

CHARLESWORTH, J.K., 1957, The Quaternary Era, Oliver and
    Boyd:Edinburgh

CHARPENTIER, J., 1835, Notice sur la cause probable du
    transport des Bloc erratiques de la Suisse, Ann. Mines
    (Paris), 8, 219-36

CHIKISHEV, A.G. (ed.), 1973, Landscape Indicators; new
    techniques in Geology and Geography (trs. FITZSIMMONS,
    J.P.), Consultants Bureau: New York (Russian ed. 1970)

CHISHOLM, M., 1967, General systems theory and geography,
    IBGT, 42, 45-52

CHORLEY, R.J., 1959, The shape of drumlins, J. Glaciol., 3,
    339-44

---, 1962, Geomorphology and general systems theory, USGSPP,
    500-B

---, 1963, The diastrophic background to twentieth-century
    geomorphological thought, GSAB, 74, 953-70

---, 1965, A re-evaluation of the geomorphic system of
    W.M.Davis, in CHORLEY, R.J. & HAGGETT, P., Frontiers in
    Geographical Teaching, Arnold:London, 21-38

---, 1978, Bases for theory in geomorphology, in EMBLETON,
    C., et al. 1978, 1-13

CHORLEY, R.J. & BECKINSALE, R.P., 1980, G.K.Gilbert's
    geomorphology, in YOCHELSON, E.L., (ed.), 1980, 129-142

CHORLEY, R.J., BECKINSALE, R.P. & DUNN, A.J., 1973, The
    History of the Study of Landforms, vol. 2, The life and
    work of W.M.Davis, Methuen:London

CHORLEY, R.J., DUNN, A.J. & BECKINSALE, R.P., 1964, The
    History of the Study of Landforms, vol. 1, Geomorphology
    before Davis, Methuen:London

CHORLEY, R.J. & KENNEDY, B.A., 1971, Physical Geography; a
    systems approach, Prentice Hall:London

CHURCH, M., 1980, Records of recent geomorphological events,
    in CULLINGFORD, R.A. et al. (eds), 1980, 13-30

CLARK, J.A. & LINGLE, C.S., 1979, Predicted relative sealevel
    changes (18,000 BP to Present), caused by Late-Glacial
    retreat of the Antarctic Ice Sheet, QR, 11, 279-298

CLARKE, J.I., 1966, Morphometry from Maps, in DURY, G.H.
    (ed.), 1966, Essays in Geomorphology, Heinemann:London,
    235-274

CLAYTON, K.M., 1970a, The problem of field evidence in

geomorphology, in OSBORNE, R.H. et al. (eds),
Geographical essays in honour of K.C.Edwards, Dept. of
Geography:Nottingham
---, 1970b, Publication and communication in Geography,
Geographica Polonica, 18, 13-20
CLOW, A. & CLOW, N.L., 1947, Dr James Hutton and the
manufacture of Sal Ammoniac, Nature, 159, 425
COATES, D.R. (ed.), 1973, Coastal Geomorphology, State Univ.
New York:Binghamton
COATES, D.R. & VITEK, J.D. (eds), 1980, Thresholds in
Geomorphology, George Allen & Unwin:London
CONEYBEARE, W., 1830/1, Examination of those phenomena of
geology which seem to bear most directly on theoretical
speculations, Phil. Mag., 8ns, 359-62, 401-6, 9ns, 19-23,
111-17, 188-97, 258-70
COOKE, R.U. & DOORNKAMP, J., 1974, Geomorphology and
Environmental Management, Arnold:London
COOKE, R.U. & REEVES, R.W., 1976, Arroyos and environmental
change in the American South-West, OUP:Oxford
CORBEL, J., 1957, Les Karsts du Nord-Ouest de l'Europe:
Inst. des etudes Rhodianiennes de L'Universite de Lyons,
Mem. et Doc., 12, 1-544
---, 1959, Erosion en terrain calcaire, Annales de
Geographie, 68, 97-120
CORNAGLIA, P., 1889, Delle Spiaggie, Accad. Naz. Lincei Atti.
Cl. Sci. Fis. Mat. et Nat., Mem 5, (4), 284-304, (Eng.
trs. in FISHER, J.S. & DOLAN, R. (eds), 1977, Benchmark
Papers in Geology, 39, Dowden, Hutchinson & Ross
Inc.:Stroudsburg (Penn.), 11-26
CORNISH, V., 1898, On sea beaches and sand banks, GJ, 11,
528-543, 628-651
COTTON, C.A., 1941, Landscape as developed by the processes
of normal erosion, Whitcombe & Tombs Ltd.:Christchurch
(NZ)
---, 1942, Climatic accidents in Landscape-making, Whitcombe
and Tombs Ltd.:Christchurch (NZ)
---, 1944, Volcanoes as landscape forms, Whitcombe & Tombs
Ltd.:Christchurch (NZ)
---, 1958, Alternating morphogenetic systems, GM, 95, 125-136
---, 1961, A theory of savannah planation, Geography, 46,
89-101
CRAIG, R., 1982, The ergodic principle in erosional models,
in THORN, C.E., (ed.), 1982, 81-116
CRAIG, R. & CRAFT, J.L., 1980, Applied Geomorphology, George
Allen & Unwin:London
CRICKMAY, C.H., 1933, The later stages of the cycle of
erosion, GM, 70, 337-347
---, 1974, The work of Rivers, MacMillan:London
---, 1975, The hypothesis of unequal activity, in MELHORN,
W.N. & FLEMAL, R.C. (eds), 1975, 103-110
CROLL, J., 1864, On the physical cause of the change of

climate during geological epochs, Phil. Mag., 28, 121-137

---, 1867, On the excentricity of the earth's orbit and its physical relations to the Glacial Period, Phil. Mag., 33, 119-31

---, 1875, Climate and Time, Dalby & Ibister:London

CULLING, W.E.H., 1957, Multicyclic streams and the equilibrium theory of grade, JG, 65, 259-74

---, 1960, Analytic theory of erosion, JG, 68, 336-44

---, 1963, Soil creep and the development of hillside slopes, JG, 71, 127-61

---, 1965, Theory of erosion of soil covered slopes, JG, 73, 230-54

---, 1983, Steady state distributions in the measurement of soil creep, Geogr. Analysis, 15, 212-230

CULLINGFORD, R.A., DAVIDSON, D.A. & LEWIN, J. (eds), 1980, Timescales in Geomorphology, John Wiley:New York

CUNNINGHAM, F., 1977a, The Revolution in Landscape Science, Tantalus Research Ltd:Vancouver

---, 1977b, Lyell and Uniformitarianism, Canadian Geogr., 21, 164-174

CURWEN, E.C., 1940, The journal of Gideon Mantell, Surgeon and Geologist, OUP:London

CUVIER, G., 1815, Essay on the Theory of the Earth, Paris (Eng. trs. KERR, E., notes by Prof. Jameson, W. Blackwood:Edinburgh (2nd ed.), later ed. 1828)

CVIJIC, J., 1893, Das Karstphanomen, Geographisce Abhandlungen herausgegeben von A. PENCK, 5(3), 218-329

---, 1960, La geographie des terrains calcaires (French trs. by DE MARTONNE, E.,), Serbe. Acad. Scio. Arts. Mon., 5 (341)

DALY, R.A., 1910, Pleistocene Glaciation and the Coral Reef problem, AJS, 30, 297-308

---, 1934, The Changing World of the Ice Age, Yale Univ. Press:Newhaven (Conn.)

DAMPIER, W., 1964, A Shorter History of Science, World Pub. Co.:Cleveland

DANA, J., 1849a, On denudation in the Pacific, in US Expl. Exped. 1838-42, Philadelphia (rprt, 1850, AJS, 2nd ser. 9, 48-62)

---, 1849b, On the degradation of the rocks of New South Wales and formation of valleys, in US Explor. Exped. 1838-1842, Philadelphia (rprt, 1850, AJS, 2nd ser. 9, 305-34)

---, 1849c, Review of Chambers (1848) 'Ancient Sea Margins', AJS, (2nd ser) 7, 1-14, 8, 86-9

---, 1853, On Coral Reefs and Islands, Putnam:New York

---, 1863, A Manual of Geology, Ivison et al.:Philadelphia

---, 1872, Corals and Coral Islands, Dodd & Mead:New York

DARBY, H.C., 1955, The clearing of the woodland in Europe, in

THOMAS, W.L. (ed.), Man's Role in Changing the Face of the Earth, Univ. Chicago:Chicago, 183-216

DARWIN, C., 1838, On certain areas of elevation and subsidence in the Pacific and Indian oceans, as deduced from the study of coral formations, Pro. Geol. Soc., 2, 552-4

---, 1839, Observations on the Parallel Roads of Glen Roy, Phil. Trans. Roy. Soc. Pt. 1, 39-81

---, 1842, The Structure and distribution of Coral Reefs, Smith, Elder & Co.:London

---, 1849, Geology, in HERSCHEL, J. (ed.), 1849

---, 1859, On the Origin of Species, John Murray:London

DAVIDSON, D.A., 1980, Erosion in Greece during the first and second millenia BC, in CULLINGFORD, R.A., et al. (eds), 1980, 143-158

DAVIES, G.L., 1969, The Earth in Decay, MacMillan:London

DAVIES, J.L., 1959, Wave refraction and the evolution of shoreline curves, Geographical Studies, 5, 1-14

---, 1964, A morphogenic approach to world shorelines, ZfG, 8nf, 127-142

---, 1973, Geographical variations in coastal development, Hafner:New York

DAVIS, W.M., 1882, Glacial erosion, Proc. Bost. Soc. Nat. Hist., 21, 315-81

---, 1884, Gorges and waterfalls, AJS, 3rd ser. 28, 123-32

---, 1885, Geographic classification, illustrated by a study of plains, plateaus and their derivatives, Proc. Amer. Assoc. Adv. Sci., 33, 428-32

---, 1889, The rivers and valleys of Pennsylvania, NGM, 1, 183-253

---, 1896, Plains of marine and subaerial denudation, GSAB, 7, 377-98

---, 1899a, The geographical cycle, GJ, 14, 481-504

---, 1899b, The peneplain, Amer. Geologist, 23, 207-93

---, 1900, Glacial erosion in France, Switzerland and Norway, Proc. Bost. Soc. Nat. Hist., 29, 273-322

---, 1902, Base level, grade and peneplain, JG, 10, 77-111

---, 1905a, Complications of the Geographical Cycle, Rept. 18th Int. Geogr. Congress, Washington 1904, 150-63

---, 1905b, The geographical cycle in an arid climate, JG, 13, 381-407

---, 1906, The sculpture of mountains by glaciers, SGM, 22, 76-89

---, 1909, Geographical Essays (JOHNSON, D.W. (ed.)) Ginn & Co.:Boston (reprt 1954, Dover:New York)

---, 1912, Die Erklarende Beschreibung der Landformen (German trs. by RUHL, A.) Teubner:Leipzig, (2nd ed. 1924)

---, 1922, Peneplains and the geographical cycle, GSAB, 33, 587-98

---, 1928, The coral reef problem, AGS:New York

---, 1930, The origin of limestone caverns, GSAB, 41, 475-628

---, 1932, Piedmont Benchlands and primarrumpfe, GSAB, 43, 399-440

---, 1980, The Physical Geography (Geomorphology) of W.M.Davis (eds KING, P.B. & SCHUMM, S.A.). Geo Books:Norwich

DAWSON, J.W., 1893, The Canadian ice age, W.V. Dawson:Montreal

DEAN, D.R., 1975, James Hutton on Religion and Geology: The unpublished Preface to his Theory of the Earth (1788), Ann. of Sci., 32, 187-193

DE BEAUMONT, E., 1845, Lecons de geologie pratique, Paris

DEBENHAM, F., 1942, A laboratory for physical geography, GJ, 100, 223-235

DE BOER, G., 1964, Spurn Head: its History and Evolution, IBGT, 34, 71-89

DE GEER, G., 1892, Quaternary changes of level in Scandinavia, GSAB, 3, 65-68

DE LA BECHE, H.T., 1831, A geological manual, Treuttel & Wurtz, Treuttel Jun. & Richter:London (later eds 1832, 1833)

---, 1839, Report on the Geology of Cornwall, Devon and West Somerset, H.M.S.O.:London

DE LA NOE, G.D. & DE MARGERIE, E., 1888, Les Formes du Terrain, Libraire Hachette:Paris

DE LAPPARENT, A., 1896, Lecon de Geographie Physique, Paris

DE LUC, J., 1790, Lettres Physique et Morales sur L'Histoire de La Terre, Monthly Review, Tome 2

DE MAILLET, B., 1749, Telliamed: or, Discourse between an Indian Philosopher and a French Missionary, Paris, (Eng. trs. 1750)

DE MARTONNE, E., 1909, Traite de Geographie Physique, Armand Colin:Paris

---, 1913, Le climat-facteur du relief, Scientia, 339-355, (Eng. trs. in DERBYSHIRE, E. (ed.), 1973, 61-75)

---, 1940, Interpretation geographique de l'hypsometrie Francaise, Compt. Rend. de l'Acad. des Sci., 211, 378-80, 426-8

DEMEK, J. (ed.), 1972, Manual of detailed geomorphological mapping, Academia:Prague

DENTON, G.H. & HUGHES, T.J. (eds), 1981, The Last Great Ice Sheets, John Wiley:New York

DERBYSHIRE, E. (ed.), 1973, Climatic Geomorphology, MacMillan:London

DERBYSHIRE, E., GREGORY, K.J. & HAILS, J.R. (eds), 1979, Geomorphological Processes, Dawson:Folkestone

DESMAREST, N., 1775, Memoire sur le clef de epoques de Nature par les produits des volcans, Mem. de l'Institut, Tome VI, Paris

DE SMET, R., 1951, Problems de morphometrie, Bull. Soc. Belge d'Etudes Geog., 20, 111-32

DOLOMIEU, G., 1784, Memoire sur les tremblemens de la terre

de la Calabre pendant l'annee 1783, Rome 1784 (Eng. trs. in PINKERTON, J., (1808-14), A general collection of the best and most interesting voyages and travels in many parts of the world, 17 volumes, London)

DONOVAN, A. & PRENTISS, J.J. (eds), 1980, see HUTTON, J. (1749)

DOORNKAMP, J. & KING, C.A.M., 1972, Numerical Methods in Geomorphology, Arnold:London

DOUGLAS, I., 1981, Tropical geomorphology: present problem and future prospects, in EMBLETON, C. et al. (eds), 1978, 162-184

DRAKE, E.T., 1981, The Hooke imprint on the Huttonian Theory, AJS, 281, 963-973

DRAKE, E.T. & KOMAR, P.D., 1981, A comparison of the geological contribution of Nicolaus Steno and Robert Hooke, J. Geol. Education, 29, 127-134

DREIMANIS, A. et al., 1957, Heavy mineral studies in tills in Ontario and adjacent areas, J.Sedimentary Petrology, 27, 48-161

DRISCOLL, E.M., 1964, Landforms in the northern territory of Australia, in STEEL, R.W. & PROTHERO, R.M. (eds), Geographers and the Tropics, Liverpool Essays, Longmans:London, 57-80

DU BUAT, L.G., 1779 and 1786, Principes d'Hydraulique, 2 vols, Paris (3rd ed. 3 vols, 1816)

DUNCAN, J.M., 1823, Travels through part of the United States and Canada in 1818 and 1819, Glasgow

DUNN, J.R. & HUDEC, P.P., 1966, Water, clay and rock soundness, The Ohio Journal of Science, 66, 153-68

DURY, G.H., 1953, A glacial breach in the Northwestern Highlands, SGM, 69, 106-117

---, 1954, Contributions to a general theory of meandering valleys, AJS, 252, 193-224

---, 1959, The face of the earth, Penguin:Harmondsworth

---, 1964a, Principles of underfit streams, USGSPP, 452-A

---, 1964b, Subsurface explorations and chronology of underfit streams, USGSPP, 452-B

---, 1965, Theoretical implications of underfit streams, USGSPP, 452-C

---, 1977, Underfit streams, retrospect, perspect and prospect, in GREGORY, K.J. (ed.), 1977a, 281-293

---, 1980, Neocatastrophism, a further look, Prog. in Phys. Geog., 4, 391-413

---, 1982, Review of JONES, D.K.C. (ed.) 1980, Prog. in Phys. Geog., 6, 140-144

DU TOIT, A., 1937, Our Wandering Continents, Oliver and Boyd:Edinburgh

DUTTON, C., 1889, On some of the greater problems of physical geology, Bull. Phil. Soc. Wash., 11, 51-64

EATON, A., 1824, A geological and agricultural survey of the district adjoining the Erie Canal, in the state of New York, Packard and van Beuthuysen:Albany

EDWARDS, A.B., 1951, Wave action in shore platform formation, GM, 88, 41-49

ELLENBERGER, F., 1975, A l'aube de la geologie moderne: Henri Gautier (1660-1737), Histoire et Nature, 7, 1-58, et 9,10, 1-149

---, 1980, De l'influence de l'environment sur les concepts: l'example des theories geodynamique au XVIIIe siecle en France, Rev. Hist. Sci., 33, 33-68

EMBLETON, C. (ed.), 1973, Climatic geomorphology, MacMillan:London

EMBLETON, C., BRUNSDEN, D. & JONES, D.K.C, (eds), 1978, Geomorphology: Present problems and future prospects, OUP:Oxford

EMBLETON, C. & KING, C.A.M., 1968, Glacial and peri-glacial geomorphology, Arnold:London

EMBLETON, C. & THORNES, J.B., 1979, Process in Geomorphology, John Wiley:New York

ESMARK, J., 1827, Remarks tending to explain the Geological History of the Earth, ENPJ, 2, 107-121, (rprt from Christiana Journal 1826)

EVANS, L., 1749 & 1752, A Map of Pensilvania, New Jersey, New York, ..., Philadelphia

EVANS, O.F., 1942, The origin of spits, bars and related structures, JG, 50, 846-865

EYLES, R.J., 1966, Stream representation on Malayan maps, J. Tropical Geography, 22, 1-9

EYLES, V.A., 1948, Note on the original publication of Hutton's Theory of the Earth, and on subsequent forms in which it was issued. Proc. Roy. Soc. Edinburgh, 63 B(4), 377-86

---, 1955, A bibliographic note on the earliest printed version of James Hutton's Theory of the Earth, its form and date of publication, J. Soc. Biblphy. nat. Hist., 3(2), 105-8

FAIR, T.D.J., 1947, Slope form and development in the interior of Natal, Trans. Geol. Soc. S. Afr., 50, 105-20

---, 1948, Slope form and development in the coastal hinterland of Natal, Trans. Geol. Soc. S. Afr., 51, 37-53

FAIRBRIDGE, R.W., 1950a, Landslide patterns on oceanic volcanoes and atolls, GJ, 115, 84-88

---, 1950b, Recent and Pleistocene coral reefs in Australia, JG, 58, 330-401

---, 1961, Eustatic changes in sea level, Physics and Chemistry of the Earth, 4, 99-185

---, 1973, Glaciation and plate migration, in TARLING, D.H. & RUNCORN, S.K. (eds), Implications of Continental Drift to

the Earth Sciences (2 Vols), Academic Press: London & New
    York, 501-512
FAIRHOLME, G., 1834, On the falls of Niagara, Phil. Mag.,
    5ns, 11-25
FAREY, J., 1815, General view of the Agriculture and Minerals
    of Derbyshire, London
FEATHERSTONEHAUGH, G.W., 1831, On the ancient drainage of
    North America, and the origin of the cataract of Niagara,
    Monthly Amer. J. of Geol. and nat. Sci., 1, 13-21
FENNEMAN, N.M., 1936, Cyclic and non-cyclic aspects of
    erosion, Science, 83, 87-94
---, 1938, Physiography of Eastern United States, McGraw
    Hill: New York & London
FISHER, J.J., 1968, Barrier island formation: discussion,
    GSAB, 79, 1421-1426
FITTON, W.H., 1839, Review of Lyell's 'Elements of Geology',
    Edin. Review, 69, 406-66
FLEMING, S., 1854, Toronto Harbour- its formation and
    preservation, Canadian Journal, 2, 103-107, 223-230,
    supp. to same, 15-29
---, 1861, Notes on the Davenport gravel drift, Canadian
    Journal, 6ns, 247-253
FLINT, J-J., 1980, Tributary arrangements in fluvial systems,
    AJS, 280, 26-45
FLINT, R.F., 1929, The stagnation and dissipation of the last
    ice sheet, GR, 19, 256-289
---, 1930, The origin of the Irish eskers, GR, 20, 615-630
---, 1938, Origin of the Cheney-Palouse scabland tract,
    Washington, GSAB, 49, 461-523
---, 1943, Growth of the North American ice sheet during the
    Wisconsin age, GSAB, 54, 325-362
---, 1947, Glacial Geology and the Pleistocene Epoch, John
    Wiley:New York
---, 1957, Glacial and Pleistocene Geology, John Wiley:New
    York
---, 1971, Glacial and Quaternary Geology, John Wiley:New
    York
FLINT, R.F. & DEMOREST, M., 1942, Glacier thinning during
    deglaciation, AJS, 240, 29-66, 113-136
FORBES, J.D., 1846, Notes on the topography and geology of
    the Cuchullin Hills, and on traces of ancient glaciers
    which they present, ENPJ, 40, 76-99
FORBES, R.J., 1957, Hydraulic engineering and sanitation, in
    A History of Technology, SINGER, C. et al. (eds), vol. 2,
    OUP:Oxford, 663-694
FORD, D.C., 1980, Threshold and limit effects in karst
    geomorphology, in COATES, D.R. & VITEK, J.D. (eds) 1980,
    345-62
FORD, D.C. et al., 1981, On the age of the existing relief in
    the southern Rocky Mountains of Canada, Arctic and Alpine
    Res., 13, 1-10

FOSTER, C.L.N. & TOPLEY, W., 1865, On the superficial
    deposits of the valley of the Medway, with remarks on the
    denudation of the Weald, QJGS, 21, 443-74
FOURNIER, F., 1960, Climat et erosion, Paris
FRONTINUS, S.J., (1961) The stratagems, and the aqueducts of
    Rome, Eng. trs. by BENNET, C.E., Harvard U.P.:Cambridge

GARDNER, J.S., 1977, Some geomorphic effects of a
    catastrophic flood on the Grand River, Ontario, Canadian
    Journal of Earth Sciences, 14, 2294-2300
GARWOOD, E.J., 1910, Features of Alpine scenery due to
    glacial protection, GJ, 36, 310-39
GAUTIER, H., 1721, Nouvelles conjectures sur le Globe de la
    Terre, Paris
GEIKIE, A., 1863, On the phenomena of the Glacial Drift of
    Scotland, Trans. Geol. Soc. Glasgow, 1(2), 1-190
---, 1865, Scenery of Scotland viewed in connexion with its
    physical geology, MacMillan:London (later eds, 1887,
    1901)
---, 1868, On denudation now in progress, GM, 5, 249-54
---, 1868/9, On modern denudation, Trans. Geol. Soc. Glasgow,
    3, 153-190
---, 1879, Outlines of Field Geology, MacMillan:London
---, 1880, Rock weathering measured by the decay of
    tombstones, Royal Society of Edinburgh, Proceedings, 10,
    518-32
---, 1882a, A textbook of Geology, MacMillan:London
---, 1882b, Geological Sketches at home and abroad,
    MacMillan:London
---, 1905, The founders of Geology, MacMillan:London
GEIKIE, J., 1874, The Great Ice Age, Dalby & Ibister:London,
    (later eds 1877, 1894)
---, 1891, The scientific results of Dr Nansens' expedition,
    in GEIKIE, J., 1893, 382-392
---, 1892, The geographical development of coast-lines (Pres.
    addr. to Brit. Assoc., Edin. 1892), in GEIKIE, J., 1893,
    393-428
---, 1893, Fragments of Earth Lore, Bartholomew:Edinburgh
---, 1898, Earth Sculpture, John Murray:London
GENERELLI, G.C., 1719, Dissertazione de' crostacei, e dell'
    altre produzioni che sono ne monti, Milan
GILBERT, G.K., 1877, Report on the Geology of the Henry
    Mountains, US Geog. and Geol. Surv.:Washington
---, 1885, The topographic features of lake shores, USGS 5th
    Ann. Rept (1883-1884)
---, 1890, Lake Bonneville, USGS Monograph, vol. 1, 23-65
---, 1895, Niagara Falls and their history, NGS Monograph 1,
    (7)
---, 1907, The rate of recession of Niagara Falls, USGS Bull.
    306

---, 1914, The transportation of debris by running water,
     USGSPP, 86
---, 1917, Hydraulic mining debris in the Sierra Nevada,
     USGSPP, 105
GILLULY, J., 1949, The distribution of mountain building in
     geologic time, GSAB, 60, 561-90
GLEN, J.W., 1952, Experiments on the deformation of ice, J.
     Glaciol., 2, 111-14
---, 1958, The flow law of ice, Int. Ass. scient. Hydr., 47,
     171-83
GLINKA, K.D., 1927, The great soil groups of the world, (trs.
     MARBUT, C.F.), Edward Brothers:Ann Arbor
GODWIN-AUSTEN, R., 1855, On land surfaces beneath the drift
     gravel, QJGS, 11, 112-119
GOULD, S.J., 1970, History versus prophecy: discussion with
     J.W.Harrington (with reply by Harrington), AJS, 268,
     187-9
GRAF, W.L., 1979a, Rapids in Canyon Rivers, JG, 87, 533-51
---, 1979b, Catastrophe Theory as a model for change in
     fluvial systems, in RHODES, D.D. & WILLIAMS, G.P. (eds),
     1979, Adjustments of the Fluvial System,
     Kendall/Hunt:Dubuque 13-32
GREENHOUGH, G.B., 1819, A critical examination of the first
     principles of Geology, Longman:London (rprt, 1978, Arno
     Press:New York)
---, 1820, Geological map of England and Wales, London
GREENWOOD, G., 1853, The Tree Lifter, (2nd ed), London
---, 1857, Rain and Rivers, London
GREGORY, K.J. (ed.), 1977a, River Channel Changes, John
     Wiley:New York
---, 1977b, Stream network volume: an index of channel
     morphometry, GSAB, 88, 1075-1080
---, 1978, Fluvial processes in British Basins, in EMBLETON,
     C. et al. (eds), 1978, 40-72
GREGORY, K.J. & BROWN, E.H., 1966, Data processing and the
     study of landforms, ZfG, 10nf, 237-263
GREGORY, K.J. & WALLING, D., 1973, Drainage basin form and
     process, Arnold:London
GREGORY, S., 1962, The raised beaches of the Peninsula area
     of Sierra Leone, IBGT, 31, 15-22
GRIGGS, D.T., 1936, The factor of fatigue in rock
     exfoliation, JG, 44, 781-96
GROVE, A.T., 1958, The ancient ergs of Hausaland and similar
     formations on the south side of the Sahara, GJ, 124,
     526-33
---, 1959, A note on the former extent of Lake Chad, GJ, 125,
     465-67
---, 1960, Geomorphology of the Tibesti region, with special
     reference to western Tibesti, GJ, 126, 18-31
---, 1969, Landforms and climatic change in the Kalahari and
     Ngamiland, GJ, 135, 191-212

GROVE, A.T. & PULLAN, R.A., 1963, Some aspects of the
    Pleistocene paleogeography of the Chad Basin, in HOWELL,
    F.C. & BOURLIERE, F. (eds),'African Ecology and Human
    Evolution,' Methuen & Co:London, 230-245
GROVE, A.T. & WARREN, A., 1968, Quaternary land forms and
    climate on the south side of the Sahara, GJ, 31, 240-263
GROVE, J.M., 1966, The Little Ice Age in the Massif of Mont
    Blanc, IBGT, 40, 129-143
GUETTARD, J.E., 1752, Memoire sur quelques montagnes de la
    France qui ont ete des volcans. Mem. de l'Academie Royale
    des Sciences, 1-8, 27-59
GUILCHER, A., 1954, Morphologie littorale et sous-marine,
    Paris, (Eng. trs. SPARKS, B.W. & KNEESE, R.H.W., 1958,
    Methuen:London)
GUILCHER, A. & KING., C.A.M., 1962, Spits, tombolos and tidal
    marshes in Connemara and west Kerry, Ireland, Proc. Roy.
    Irish Acad., 61(B), 283-338
GULLIVER, F.P., 1896, Cuspate Forelands, GSAB, 7, 399-422
---, 1897, Dungeness Foreland, GJ, 9, 536-46
---, 1899, Shoreline topography, Proc. Amer. Acad. Arts &
    Sci., 34, 151-258

HACK, J.T., 1941, Dunes of the western Navajo County, GR, 31,
    240-263
---, 1960, Interpretation of erosional topography in humid
    temperate regions, AJS, 258A, 80-97
---, 1975, Dynamic equilibrium and landscape evolution, in
    MELHORN, W.N. & FLEMAL, R.C. (eds), 1975, 87-102
HAILS, J.R. (ed.), 1977, Applied Geomorphology,
    Elsevier:Amsterdam
HALL, B., 1829, Forty etchings, from sketches made with the
    camera lucida in North America in 1827 and 1828, London
HALL, J., 1805, Experiments on Whinstone and Lava, RSET, 5(pt
    1), 43-75
---, 1812, Account of a series of experiments, shewing the
    effects of compression in modifying the action of heat,
    RSET, 6, 71-185
---, 1815, On the Revolutions of the Earth's surface, RSET,
    8, 139-67, 169-212
HALL, J., 1842, Niagara Falls - their physical changes and
    the geology and topography of the surrounding country,
    Bost. J. Nat. Hist., 4, 106-134
---, 1843, Geology of New-York, Albany
HALLAM, A., 1973, A revolution in the Earth Sciences,
    Clarendon Press:Oxford
HAMELIN, L., 1961, Periglaciaire du Canada: idees nouvelles
    et perspective globales, Cah. de Geogr. de Quebec, 10,
    141-203
HANSEN, B.L. & LANGWAY, C.C., 1966, Deep core drilling in ice
    and core analysis at Camp Century, Greenland, 1961-66,

Antarctic Journal of the U.S., 1, 207-8

HARDY, J.R., 1964, The movement of beach material and wave action near Blakeney Point, Norfolk, IBGT, 34, 53-69

HARRINGTON, J.W., 1967, The first, first principles of geology, AJS, 265, 449-461

---, 1969, The pre-natal roots of geology - a study in the history of ideas, AJS, 267, 592-97

---, 1970, Ontology of geologic reasoning with a rationale for evaluating historical contributions, AJS, 269, 295-303

HAYDEN, F.V., 1862, Some remarks in regard to the period of elevation of those ranges of the Rocky Mountains, near the sources of the Missouri River and its tributaries, AJS, 2nd ser 33, 305-13

HERODOTUS, (1920/2), The Histories, Eng. trs. by GODLEY, A.D., Heinemann:London

HERSCHEL, J. (ed.), 1849, A manual of Scientific Enquiry, John Murray: London

HETTNER, A., 1921, Die Oberflachenformen des Festlandes, Teubner:Leipzig, (2nd ed. 1928, as Eng. trs., 1972, TILLEY, P., The Surface Features of the Earth, MacMillan:London)

HEY, R.D., 1979a, Dynamic process-response model of river channel development, ESP, 4, 59-72

---, 1979b, Causal and functional relations in fluvial geomorphology, ESP, 4, 1769-182

HICKIN, E.J. & NANSON, G.C., 1975, The character of channel migration on the Beatton River, Northeast British Columbia, Canada, GSAB, 86, 487-494

HIGGINS, C.G., 1975, Theories of landscape development: A perspective, in MELHORN, W.N & FLEMAL, R.C. (eds), 1975, 1-28

HILLS, E.S., 1949, Shore platforms, GM, 86, 137-152

HITCHCOCK, E., 1841, First anniversary address before the Association of American Geologists, AJS, 41, 232-75

HJULSTROM, F., 1932, Das Transportvermogen der Flusse und die Bestimmung des Erosionsbetrages, Geogr. Annal.,14, 244-258

---, 1935, Studies of the morphological activity of rivers illustrated by the River Fyris, Uppsala Univ. Geol. Inst. Bull., 25, 221-527

HOBBS, W.H., 1910, The cycle of mountain glaciation, GJ, 35, 267-386

---, 1911, Characteristics of Existing Glaciers, MacMillan:New York

---, 1943, Discovery in eastern Washington of a new lobe of the continental Pleistocene glacier, Science, 98, 227-230

---, 1967, The glacial history of the Scabland and Okanogan lobes, Cordilleran, Continental glacier, J.W.Edwards:Ann Arbor

HODGE, E.T., 1934, Origin of the Washington scabland,

Northwest Science, 8, 4-11

HOFF, K.E.A. von, 1822-34, Geschichte du durch Uberlieferung nachgewiesenen naturlichen Veranderingen der Erdoberflache, Gotha (3 volumes)

HOLLIN, J.T., 1962, On the glacial history of Antarctica, J. of Glaciol., 4, 173-195

HOLLINGWORTH, S.E., 1952, A note on the use of marginal drainage in the recognition of unglaciated enclaves, J. of Glaciol, 2, 107-108

HOLMES, C.D., 1941, Till fabric, GSAB, 52, 1299-1354

HOMER, (1966), The Iliad, Eng. trs. by RIEN, E.V., Penguin:Harmondsworth

HOOKE, R., 1705, Lectures and discourses of earthquakes and subterraneous eruptions, R. Waller:London (rprt, 1978, Arno Press:New York)

HOPKINS, W., 1842, On the elevation and denudation of the district of the lakes of Cumberland and Westmoreland, Proc. Geol. Soc., 3, 757-66

---, 1844, On the transport of erratic blocks, Trans. Cam. Phil. Soc., 8(2), 220-40

HOPPE, G., 1957, Problems of glacial morphology and the ice age, Geogr. Annal., 39, 1-18

HORTON, R.E., 1945, Erosional development of streams and their drainage basins; hydrophysical approach to quantitative morphology, GSAB, 56, 275-370

HOUGH, J.L., 1958, Geology of the Great Lakes, University of Illinois:Urbana

HOWARD, A.D. & DOLAN, R., 1981, Geomorphology of the Colorado River in the Grand Canyon, JG, 89, 269-298

HSU, K.J., 1972, When the Mediterranean dried up, Sci. Amer., 227(6), 27-36

HULL, E., 1878, The Physical Geology and Geography of Ireland, E. Stanford:London

HUMBOLDT, A. von, 1849, Cosmos: A sketch of a physical description of the Universe, 3 vols (Eng. trs. OTTE, E.C.) H.G.Bohm:London

HUMPHREYS, A.A. & ABBOTT, H.L., 1861, Report on the physics and hydraulics of the Mississippi River, Corps of Topogr. Eng., Prof. Paper, 4, Washington

HUTTON, J., 1749, Dissertatio Physico-Medico Inauguralis de Sanguine et Circulatione Microcosmi, Leiden, (rprt and Eng. trs., DONOVAN, A., and PRENTISS, J.J., 1980, Amer. Phil. Soc.:Philadelphia)

---, (1785), Abstract of a dissertation read in the Royal Society of Edinburgh upon the Seventh of March, and the Fourth of April M,DCC,LXXXV, concerning the System of the Earth, its duration and Stability, (date, place of publication and printer all unknown, see EYLES, 1955, assumed 1785, Edin.)

---, 1788, Theory of the Earth, or an Investigation of the Laws observable in the Composition, Dissolution and

Restoration of Land upon the globe, RSET, 1(2), 209-304
(with only minor amendments and the addition of footnotes
this forms Chapter 1 of HUTTON, J., 1795)
---, 1795, The Theory of the Earth, 2 vols, William
Creech:Edinburgh (rprt, 1959, Wheldon & Wesley and
Hafner:London & New York)
---. 1899, The Theory of the Earth, Vol III, edited by Sir
ARCHIBALD GEIKIE, London
---, (1978), James Hutton's Theory of the Earth: The Lost
Drawings, Scottish Academic Press:Edinburgh
HUXLEY, T.H., 1877, Physiography: an introduction to the
study of nature, MacMillan:London

INMAN, D.L. & NORDSTROM, C.E., 1971, On the tectonic and
morphologic classification of coasts, JG, 79, 1-21
ISACHENKO, A.G., 1973, Principles of Landscape Science and
Physical-Geographic Regionalization (trs. ZATORSKI, R.J.)
Melbourne University Press:Melbourne
IVES, J.D., 1973, Arctic and alpine geomorphology- a review
of current outlook and notable gaps in knowledge, in
FAHEY, B.D. and THOMPSON, R.D. (eds), Research in Polar
and Alpine Geomorphology, Geo Abstracts:Norwich, 1-10

JAMIESON, T.F., 1865, On the history of the last glacial
changes in Scotland, QJGS, 21, 161-203
---, 1882, On the cause of the depression and re-elevation of
the land during the Glacial Period, GMns decade 2,
9(9-10), 400-407, 457-466
JEFFREY, F., 1822, Biographical account of the late Professor
Playfair, in PLAYFAIR, J.G. (ed.), 1822
JOHNSON, D.W., 1919, Shore Processes and Shoreline
Development, John Wiley:New York
---, 1931, Planes of lateral corrasion, Science, 73, 174-177
---, 1940, Walther Penck's contribution to geomorphology,
comments by D.W.Johnson, AAAG, 30, 228-232
JOLIFFE, I.P., 1961, The use of tracers to study beach
movements; the measurement of littoral drift by a
fluorescent technique, Rev. de Geom. Dynamique, 12, 81-98
JONES, D.K.C. (ed.), 1980, The shaping of southern Britain,
IBG Spec. Pub. 11
JUDSON, J.T., 1975, Evolution of Appalachian topography, in
MELHORN, W.N & FLEMAL, R.C. (eds), 1975, 29-44
JUKES, J.B., 1847, Narrative of the Surveying voyage of HMS
'Fly', London
---, 1857, Student's Manual of Geology, Black:Edinburgh
---, 1862, On the mode of formation of some of the
River-Valleys in the south of Ireland, QJGS, 18, 378-403

KAYE, C.A., 1957, Military geology in the United States
    sector of the European Theatre of operations during World
    War II, GSAB, 68, 47-54
KENDALL, P.F., 1902, A system of glacier lakes in the
    Cleveland Hills, QJGS, 58, 471-571
KEULEGAN, G.H., 1948, An experimental study of submarine sand
    bars, BEB Tech. Rept, 3, 1-40
KEYES, C.R., 1912, Deflative scheme of the geographical cycle
    in n arid climate, GSAB, 23, 537-562
KIDSON, C., 1950, Dawlish Warren: a study of the evolution of
    the sand spits across the mouth of the River Exe in
    Devon, IBGT, 16, 67-80
---, 1959, The uses and limitations of vegetation in shore
    stabilization, Geography, 44, 241-250
---, 1960, The shingle complexes of Bridgewater Bay. IBGT,
    28, 75-87
---, 1963, The growth of sand and shingle spits across
    estuaries, ZfG 7nf, 1-22
KIDSON, C. & CARR, A.P., 1959, The movement of shingle over
    the sea bed close inshore, GJ, 125, 380-389
---, 1960, Beach drift experiments at Bridgewater Bay,
    Somerset, Proc. Bristol Nat. Soc., 30, 163-180
---, 1962, Marking beach material for tracing experiments, J.
    Hydr. Div. Amer. Soc. civil Eng., 88, HY 4 43-60 (paper
    3189)
KIDSON, C. et al., 1956, Drift experiments with pebbles,
    Nature, 178, 257
---, 1958, Further experiments using radioactive methods to
    detect the movement of shingle over the sea bed and
    alongshore GJ, 124, 210-218
---, 1962, A trial of the potential value of aqua-lung diving
    to coastal physiography on British coasts, GJ, 128, 49-53
KING, C.A.M., 1951, The depth of disturbance of sand in sea
    beaches by waves, J. Sed. Pet., 21, 131-140
---, 1953, The relation between wave incidence, wind
    direction and beach changes at Marsden Bay, Co. Durham,
    IBGT, 19, 13-23
---, 1959, Beaches and Coasts, Arnold:London (2nd ed. 1972)
---, 1978, Coastal geomorphology in the United Kingdom, in
    EMBLETON, E. et al., 1978, 224-250
KING, C.A.M. & McCULLAGH, M.J., 1971, A simulation model of a
    complex recurved spit, JG, 79, 22-37
KING, C.A.M. & WILLIAMS, W.W., 1949, The formation of sand
    bars by wave action, GJ, 113, 70-85
KING, L.C., 1942, South Africa Scenery, Oliver &
    Boyd:Edinburgh
---, 1953, Canons of landscape evolution, GSAB, 64, 721-52
---, 1956, Pediplanation and isostasy: an example from South
    Africa, QJGS, 111, 353-9
---, 1962, Morphology of the Earth, Oliver and Boyd:Edinburgh
    (2nd ed. 1967)

KING, P.B. & SCHUMM, S.A. (eds), 1980, The physical geography
    (geomorphology) of W.M.Davis, Geo Books:Norwich
KING, W.B.R., 1951, Influence of geology on Military
    operations in N.W.Europe, Adv. Sci., 30, 131-7
KIRCHER, A., 1664, Mundus Subterraneus, Amsterdam
KIRWAN, R., 1799, Geological Essays, D.Bremner: London (rprt
    1978, Arno Press:New York)
KLIMASZEWSKI, M., 1956, The principles of the
    geomorphological survey of Poland, Przeglad Geograficzny,
    28 Suppl., 32-40
---, 1961, The problems of the geomorphological and
    hydrographical map on the example of the Upper Silesian
    Industrial District, Geogr. Studies, No 25, Polish
    Academy of Sciences:Warsaw
KLINEFELTER, W., 1971, Lewis Evans and his maps, rprt &
    commentary, Amer. Phil. Soc.:Philadelphia
KOEPPEN, W., 1901, Versuch einer Klassification der Klimate
    vorzugsweise nach ihren Beziehungen zur Pflanzenwelt,
    Geogr. Zeit., 6, 593
KOMAR, P.D., 1976, Beach processes and sedimentation,
    Prentice Hall:New Jersey

LADD, H.S. et al., 1950, Organic growth and sedimentation on
    an atoll, JG, 58, 410-425
LAMARCK, J.B., 1802, Hydrogeologie, Paris (Eng. trs. and
    intro., CAROZZI, A.V., 1964, Univ. Illinois Press:Urbana)
LANGBEIN, W.B. & LEOPOLD, L.B., 1966, River meanders - theory
    of minimum variance, USGSPP, 422H
LAGRANGE, J.L., 1811-15, Mecanique analytique, 2 vols
    Nouvelle edition, Paris
LA PLACE, P.S., 1796, Systeme du Monde, Cercle Sociale:Paris
LAUTRIDON, J.P. & OZOUF, J.C., 1982, Experimental frost
    shattering: fifteen years of research at the centre de
    Geomorphologie de C.N.R.S., Prog. in Physical Geography,
    6, 215-232
LAWSON, A.C., 1915, Epigene profiles of the desert, Univ.
    Calif. Pub. in Geol., Bull. 9, 23-48
---, 1936, The Sierra Nevada in the light of isostasy, GSAB,
    47, 1691-1712
---, 1948, Some further implications of the doctrine of
    isostasy, GSAB, 59, 199-210
LEAKEY, L.S.B., 1931, East African Lakes, GJ, 77, 497-514
---, Stone Age Africa: An outline of Prehistory in Africa,
    Oxford University Press:London
LEE, H.D.P., 1952, see ARISTOTLE, (1952)
LEE, W.T., 1922, The face of the earth as seen from the air,
    AGS Spec. Pub. 4
LEGENDRE, A.M., 1799, Method pour determiner la longueur
    exacte du quart du meridien, d'apres les observations
    faites pour la mesure d l'arc compris entre Dunkerque et

Barcelonne, Paris

LEHMANN, H., 1936, Morphologische Studien auf Java, Geogr. Abhandlungen, Reihe 3, 9, 1-114

---, 1954, Der tropische Kegelkarst der verschiedenen Klimazonen, Erdkunde, 8, 130-139

---, 1964, States and tasks of research on karst phenomena, Erdkunde, 16, 81-38

LEIBNITZ, G.W., 1719, Protogaea, I.G.SchmitGottingen

LEIGHLY, J.B., 1932, Towards a theory of the morphologic significance of turbulence in the flow of water in streams, University of California Publications in Geography: Berkeley, 6(1), 1-22

---, 1934, Turbulence and the transportation of rock debris by streams, GR, 24, 453-64

---, 1940, Walther Penck's contribution to geomorphology, comments by J. Leighly, AAAG, 30, 323-32

LEOPOLD, L.B., 1953, Downstream change of velocity in rivers, AJS, 251, 606-24

---, 1962, The vigil network, Int. Assoc. Sci. Hydr. Bull., 7, 5-9

LEOPOLD, L.B., EMMETT, W.W. & MYRICK, R.W., 1966, Channel and hillslope processes in a semi-arid area, New Mexico, USGSPP, 352-G

LEOPOLD, L.B. & LANGBEIN, W.B., 1962, The concept of entropy in landscape evolution, USGSPP, 500-A

---, 1963, Association and indeterminacy in geomorphology, in ALBRITTON, C.C. (ed.), 1963, 184-92

LEOPOLD, L.B. & MADDOCK, J., 1953, The hydraulic geometry of stream channels and some physiographic implications, USGSPP, 252

LEOPOLD, L.B. & WOLMAN, M.G., 1957, River channel patterns: braided, meandering and straight, USGSPP, 282-B, 39-85

---, 1960, River Meanders, GSAB, 71, 769-94

LEOPOLD, L.B., WOLMAN, M.G. & MILLER, J.P., 1964, Fluvial Processes in Geomorphology, W.H.Freeman:San Francisco

LEWIS, W.V., 1931, The effect of wave incidence on the configuration of a shingle beach, GJ, 78, 129-148

---, 1932, The formation of Dungeness Foreland, GJ, 80, 309-324

---, 1938a, The evolution of shoreline curves, PGA, 49, 249-265

---, 1938b, A meltwater hypothesis of cirque formation, GM, 75, 249

---, 1954, Pressure release and glacial erosion, J. Glaciology, 2, 417-22

---, 1960 (ed.), Norwegian Cirque Glaciers, Royal Geographical Society Research Series #4

LEWIS, W.V. & BALCHIN, W.G.V., 1940, Past sea-levels at Dungeness, GJ, 96, 258-285

LINTON, D.L., 1945, Interpretation of air photographs, Geography, 31, 89-97

---, 1949, Watershed breaching by ice in Scotland, IBGT, 17, 1-16
---, 1951, Problems of Scottish Scenery, SGM, 67, 65-85
---, 1955, The problem of tors, GJ, 121, 470-487
---, 1963, The forms of glacial erosion, IBGT, 33, 1-28
---, 1964, The origin of Pennine tors- an essay in analysis, ZfG, 8nf, 5-24
LOBECK, A.K., 1939, Geomorphology, An introduction to the study of landscapes, McGraw Hill:New York
LOGAN, W., 1863, Geology of Canada, Toronto
LOUGEE, R.J., 1940, Deglaciation of New England, J. Geomorphology, 3, 189-217
LOUIS, H., 1934, Glazialmorphologische Studien in den Gebirgen der Britischen Inseln, Berliner Geogr. Arbeiten, 6, 1-39
---, 1957, Rumpfflachenproblem, Erosionzyklus und Klimamorphologie, Geomorph. Studien Hrsg. v. Herbert Louis a Ing. Schaefer, 262, 9-26, (Eng. trs. in DERBYSHIRE, E. (ed.) 1973)
LOZINSKI, W., 1912, Die periglaziale Fazies der mechanischen Verwitterung, Comptes Rendus, 11th int. geol. Congr. (Stockholm 1910) 2, 1039-53
LYELL, C., 1830, The Principles of Geology, vol i-1830, vol. ii-1832, vol. iii-1833, John Murray:London (later eds 1834, 1835, 1837, 1840, 1847, 1850, 1853, 1868, 1872, 1875, rprt 1st ed. with intro. by RUDWICK, M.J.S., 1969, Johnson Reprint Corp:New York & London)
---, 1838, Elements of Geology, John Murray:London
---, 1842, On the recession of the falls of Niagara, Proc. Geol. Soc., 3, 595-662
---, 1845, Travels in North America (2 vols), John Murray:London (rprt, 1978, Arno Press:New York)
---, 1847, On the delta and alluvial deposits of the Mississippi, AJS, 2nd ser. 3, 34-39
---, 1849, A second visit to the United States (2 vols), John Murray:London (rprt, 1978, Arno Press:New York)
LYELL, C. & MURCHISON, R.I., 1829, On the excavation of valleys, ENPJ, 7, 15-48

MABBUTT, J.A., 1961, A stripped land surface in Western Australia, IBGT, 29, 101-114
MACKIN, J.H., 1948, Concept of the graded river, GSAB, 59, 463-511
---, 1963, Rational and empirical methods of investigation in geology, in ALBRITTON, C.C., (ed.), 1963, 135-63
MACKINTOSH, D., 1869, The scenery of England and Wales, Longmans:London
MACLURE, W., 1817, Observations on the Geology of the United States, Philadelphia
MALDE, H.E., 1968, The catastrophic late Pleistocene

Bonneville Flood in the Snake River plain, Idaho, USGSPP, 596

MANNERFELT, C.M., 1945, Nagra glacial-morpholgiska formelement, Geogr. Annal., 27, 1-239

---, 1949, Marginal drainage channels as indicators of the gradients of Quaternary ice caps, Geogr. Annal., 31, 194-199

MANTELL, G.A., (1819-1852), Journal or diary for the period 1819 to 1852, see CURWEN, E.C., 1940

---, 1836, Thoughts of a pebble, London

---, 1838, Wonders of Geology, London

MARKS, E.O., 1913, Notes on a portion of the Burdekin Valley with some queries as to the universal applicability of certain physiographic theories, Proc. Roy. Soc. Queensland, 24, 93-102

MARR, J.E., 1896, The waterways of the English Lakeland, GJ, 7, 602-625

---, 1900, The Scientific Study of Scenery, Methuen:London

MARTEL, P., 1744, An account of the glaciers or Ice Alps in Savoy, London

MATTHES, F., 1900, Glacial sculpture of the Bighorn Mountains, Wyoming, USGS 21st Ann. Rept (2), 173-190

McCAUSLIN, R., 1793, An account of an earthy substance found near the Falls of Niagara and vulgarly called the spray of Niagara, together with some remarks on the Falls, Amer. Phil. Soc. Trans., 3, 17-24

McCLAREN, C., 1842, The glacial theory of Prof. Agassiz, AJS, 42, 346-365

McCLAY, W., 1790, in Edgar S. McClay, Journal, Appleton:New York, entry for Feb. 1, 1790

McCONECHY, A., 1845, New Statistical Account of Scotland, (Berwick), Edinburgh

McGEE, WJ, 1888a, The geology of the head of Chesapeake Bay, 7th Annual Rept of the USGS for 1885-6: Washington, 545-646

---, 1888b, The classification of geographic form by genesis, NGM, 1, 27-36

---, 1891, The Pleistocene History of Northestern Iowa, 11th Annual Rept. of the USGS for 1889-90:Washington 189-577

---, 1893, The geology of Washington, in Int. Geol. Congr., V, Washington (1891), Comptes Rendus, 219-251

---, 1897, Sheetflood erosion, GSAB, 8, 87-112

McKEE, E.D. (ed.), 1979, Global Sand Seas, USGSSPP 1052, US Government Printing House:Washington

MEADE, R.H., 1969, Errors in using modern stream-load data to estimate natural rates of denudation, GSAB, 80, 1265-74

MELHORN, W.N. & EDGAR, D.E., 1975, The case for episodic, continental-scale erosion surfaces: a tentative geodynamic model, in MELHORN, W.N. & FLEMAL, R.C. (eds) 1975, 243-276

MELHORN, W.N. & FLEMAL, R.C. (eds), 1975, Theories of

Landform Development, George Allen & Unwin:London

MELTON, M.A., 1958a, Geometric properties of mature drainage systems and their representations in an E4 phase space, JG, 66, 25-54

---, 1958b, Correlation structure of morphometric properties of drainage systems and their controlling agents, JG, 442-60

---, 1960, Intravalley variation in slope angles related to microclimate and erosional environment, GSAB, 71, 133-44

MERRILL, G.P., 1897, A Treatise on Rocks, Rock-weathering and Soils, MacMillan:New York

MEYERHOFF, H.A., 1976, The Penckian model - with modifications, in MELHORN, W.N. & FLEMAL, R.C. (eds), 1975, 45-68

MICHELL, J., 1760, Essay on the cause and phenomena of Earthquakes, Phil. Trans., 51, 506-634

MILLER, H., 1854, The testimony of the rocks, Constable:Edinburgh

MILNE, A.A., 1928, The house at Pooh Corner, Methuen & Co.:London

MONTLOSIER, F. de, 1788, Essai sur la theorie des volcans, Paris

MORO, L., 1740, Sui crostacei et altri Corpi marini che si trovano sui Monti (On the marine bodies which are found in mountains), Naples

MORTENSEN, H., 1932, Blockenmeere und Felsburgen in dem deutschen Mittelgebirgen, Zeit. fur Gesellschaft fur Erdkunde zu Berlin, 279-287

MUIR, J., 1915, Travels in Alaska, Houghton Mifflin:Boston

MULLER, R.A. & OBERLANDER, T.O., 1978, Physical Geography Today (2nd ed.) CRM/Random House:New York

MULLER, S.W., 1945, Permafrost or permanently frozen ground and related engineering problems, USGS Special Report, Strategic Engineering Study 62, (2nd ed.)

MURCHISON, R.I., 1839, The Silurian System, John Murray:London

MURRAY, J., 1802, A comparative review of the Huttonian and Neptunian Systems of Geology, Ross & Blackwood:Edinburgh (rprt, 1978, Arno Press:New York)

NACE, R.L., 1958, Military use of geologists in World War II, GSAB, 69, 615-6

NANSEN, F. (ed), 1890, The first crossing of Greenland, (Eng. trs. GEPP, H.M.) Longman:London

NAUMANN, C.F., 1858, Lehrbuch der Geognosie, Engelmann:Leipzig

NEWBERRY, J.S., 1862, Colorado River of the West, AJS, 2nd ser. 33, 387-403

---, 1876, Geological report of the exploring expedition from Santa Fe, New Mexico, to the junction of the Grand and

Green Rivers of the Great Colorado of the West in 1859,
    Washington
NEWBIGIN, M.I. & FLETT, J.S., 1917, James Geikie: The man and
    the geologist, Oliver & Boyd:Edinburgh
NILSSON, E., 1931, Quaternary glaciation and pluvial lakes in
    British East Africa, Geogr. Annal., 13, 249-349
NORTH, F.J., 1943, Centenary of the glacial theory, PGA, 54,
    1-28
NYE, J.F., 1959, The motion of ice-sheets and glaciers, J.
    Glaciol., 2, 493-507
---, 1965, The flow of a glacier in a channel of rectangular,
    elliptic, or parabolic cross-section, J. Glaciol., 5,
    661-90

O'BRIEN, M.P., 1933, Review of the theory of turbulent flow
    and its relation to sediment transportation, Trans. Amer.
    Geophys. Union, 14th Ann. Mtg, 487-491
OBRUCHEV, V.A., 1948, Osnovnyje certy kinetiki i plastiki
    neotektoniki, Izv. Acad. Nauk. SSSR. Ser. Geol. #5
OGILVIE, A.G., 1936, The earth sculpture laboratory, GJ, 87,
    145-9
OLLIER, C.D., 1969, Weathering, Oliver and Boyd:Edinburgh
---, 1979, Evolutionary geomorphology of Australia and
    Papua-New Guinea, IBGT, 4ns, 516-539
OVID, P., (1964), Metamorphoses, Eng. trs. by INNES, M.M.,
    Penguin:Harmondsworth

PALISSY, B., 1580, Discours Admirables, de la nature des eaux
    et fonteines,..des pierres, des terres,.., Paris
PALLAS, P.S., 1777, Obersvations sur la formation des
    montagnes et les changements arrives au globe,
    particulierment de l'Empire Russe, St Petersburg
PALMER, H.R., 1834, Observations on the motions of shingle
    beaches, Phil. Trans., 124, 567-576
PALMER, J., 1956, Tor formation at the Bridestones in
    north-east Yorkshire and its significance to problems of
    valley side development and regional glaciation, IBGT,
    22, 55-71
PALMER, J. & RADLEY, J., 1961, Gritstone tors of the English
    Pennines, ZfG, 5nf, 37-52
PALMQUIST, R.C., 1975, The compatability of structure,
    lithology and geomorphic models, in MELHORN, W.N. &
    FLEMAL,. R.C (eds), 1975, 145-68
PARDEE, J.T., 1942, Unusual currents in glacial Lake
    Missoula, GSAB, 53, 1569-1600
PARRY, J.T., 1967, Geomorphology in Canada, Canadian Geog.,
    9, 280-311
PASSARGE, S., 1904, Die Kalahari, Berlin
---, 1926, Morphologie der Klimazonen oder morphologie der

Landschaftsgurtel? Petermanns Geog. Mitt., 72, 173-175
(Eng. trs. in DERBYSHIRE, E., 1973)

PATTON, P.C. & BAKER, V.R., 1976, Morphometry of floods in
small drainage basins subject to diverse
hydro-geomorphological controls, Water Resources
Research, 12, 941-52

---, 1977, Geomorphic response of central Texas stream
channels to catastrophic rainfall and runnof, in
DOEHRING, D.O. (ed.), 1977, Geomorphology and arid
regions, State Univ. New York:Binghamton, 187-217

PAUSANIAS, (1971), A guide to Greece, Eng. trs. by LEVI, P.,
Penguin:Harmondsworth

PEEL, R.F., 1949, A study of two Northhumberland spillways,
IBGT, 15, 73-89

---, 1966, The landscape in aridity, IBGT, 38, 1-23

PEGUY, Ch-P., 1948, Introduction a l'emploi des methodes
statistique en geographie physique, Rev. de Geog. Alpine,
36, 1-103

PELTIER, L.C., 1950, The geographical cycle in periglacial
regions as it is related to climatic geomorphology, AAAG,
40, 214-236

PENCK, A., 1886/7, Uber Denudation der Erboberflache, Schr.
zur Verbr. naturwiss. Kenntnisse, 27, Wien

---, 1894, Morphologie der Erdoberflache, 2 vols,
Fngelhorn:Stuttgart

---, 1910, Versuch einer Klimaklassification auf
physiographische Grundlage, Preussen Akad. der Wissen,
Sitz. der phys.-math., kl. 12, 236-246 (Eng. trs. in
DERBYSHIRE, E., 1973, 51-59)

---, 1926, Reply to the review by I. Bowman, GR, 16, 350-1

PENCK, A. & BRUCKNER, E., 1909, Die Alpen im Eiszeitalter,
Tauchnitz:Leipzig

PENCK, W., 1924, Die morphologische Analyse: Ein Kapitel der
physicalischen Geologie, Geogr.Abhandlungungen, 2.Reihe,
Heft 2, Stuttgart (Eng. trs. by CZECH, H. & BOSWELL,
K.C., 1953, Morphological Analysis of Landforms,
MacMillan:London)

---, 1925, Die piedmontflachen des sudlichen Schwarzaldes,
Zeit. Gessel. Erdk. Berlin,83-108

PHILLIPS, J., 1831, On some effects of the atmosphere in
wasting the surfaces of buildings and rocks, Proc. Geol.
Soc., 1, 323-4

PIRSSON, L.V. & SCHUCHERT, C., 1915, A textbook of Geology,
John Wiley:New York

PLATO, (1962), Timaeous, Critias, Cleitophan, Menexemus &
Epistles, Eng. trs. by BURY, R.G., Harvard Univ,
Press:Cambridge

PLAYFAIR, J., 1802, Illustrations of the Huttonian Theory of
the Earth, William Creech:Edinburgh (French trs. by
BASSET, C.A., 1815, rprt in PLAYFAIR, J.G. (ed.) 1822,
rprt, 1964, with intro. by WHITE, G.W., Dover:New York)

---, 1805, A biographical account of the late Dr James
    Hutton, F.R.S. Edin., RSET, 5, 39-99
---, 1811, Account of a lithological survey of Schehallien,
    made in order to determine the specific gravity of the
    rocks which compose that mountain, Phil. Trans., 101,
    347-77
---, 1814, Outlines of Natural Philosophy (3rd ed. 1819)
    Constable:Edinburgh
---, 1816, ..a letter to the Royal Society of Edinburgh
    concerning the origin of the Parallel Roads of Glen Roy,
    quoted by CHAMBERS, R., 1848, p107. (I have been unable
    to trace or verify the existence of this letter, either
    in published sources or at the Royal Society of
    Edinburgh.)
---, 1820, Catalogue of the library of the late John
    Playfair, Esq., F.R.S.L. & E., Professor of Natural
    Philosophy in the University of Edinburgh. Compiled by
    John Ballantyre, auctioneer for the sale beginning
    January 20th 1820, J.Ballantyre:Edinburgh
PLAYFAIR, J.G. (ed.), 1822, The collected works of John
    Playfair (4 vols), London
PLINY, C., (1956) Natural History, Eng. trs. by RACKHAM, H.,
    Harvard Univ. Press:Cambridge
POUCHOT, F., 1790, Memoir upon the late war in North America
    (Eng. trs. 1781, later ed. by HOUGH, F.B.,
    Woodward:Roxbury (Mass))
POWELL, J.W., 1875, Exploration of the Colorado River of the
    West (1869-72), Washington (rprt, 1957, Univ. Of Chicago
    and Univ. of Cambridge)
---, 1876, Report on the geology of the eastern portion of
    the Uinta Mountains, Washington
POWNALL, T., 1776, A topographical description of such parts
    of North America as are contained in the (annexed) map of
    the middle British colonies of North America,
    J.Almon:London
PRESTWICH, J., 1852, On some effects of the Holmfirth flood,
    QJGS, 8, 225-230
PRICE, W.A., 1950, Saharan dunes and the origin of the
    longitudinal dunes - a review, GR, 40, 462-465
PRIOR, D.B., 1978, Some recent progress and problems in the
    study of mass movements in Britain, in EMBLETON, E. et
    al. (eds), 1978, 84-106
PUGH, J.C., 1956, Isostatic readjustment and the theory of
    pediplanation, QJGS, 111, 361-374
PYNE, S., 1975, The mind of Grove Karl Gilbert, in MELHORN,
    W.N. & FLEMAL, R.C. (eds), 1975, 277-298

RAMSAY, A.C., 1846, The denudation of South Wales, Mem. Geol.
    Surv. Grt. Brit., 1, 297-335, HMSO:London
---, 1859, The old glaciers of Switzerland and North Wales,

in Ball, J. (ed.), Peaks, passes and glaciers:London

---, 1862, On the glacial origin of certain Lakes in Switzerland, the Black Forest, Great Britain, Sweden, North America, and elsewhere, QJGS, 18, 185-204

---, 1863, The Physical Geology and Geography of Great Britain, Stanford, London (5th ed. 1878)

RAPP, A., 1959, Avalanche boulder tongues in Lappland, a description of a little known form of periglacial accumulation, Geogr. Annal., 41, 34-48

---, 1960a, Talus slopes and mountain walls at Tempelfjorden, Spitzbergen, Nordsk Polarinstituts Skrifter, 5, 119

---, 1960b, Recent developments of mountain slopes at Karkevagge and surroundings, northern Scandinavia, Geogr. Annal., 42, 65-200

---, 1962, Karkevagge: some recordings of mass movements in the northern Scandinavian mountains, Biuletyn Periglacjalny, 11, 287-309

RAY, J., 1793, Three Physico-Theological discourses (2nd ed.), W. Innys:London (rprt of 1713, 3rd ed., 1978, Arno Press:New York)

RECLUS, E., 1871, The Earth (Eng. trs. by WOODWARD, B.B.), Chapman & Hall:London

RECTOR, R.L., 1954, Laboratory study of the equilibrium profiles of beaches, BEB Tech. Mem. 41

REICHE, P., 1945, A survey of weathering processes and products, University of New Mexico Publications in Geology #3:Albuquerque (revised 1950, reprt 1962)

RICH, J.L., 1938, The recognition and significance of multiple erosion surfaces, GSAB, 49, 1695-1722

RICHARDSON, W., 1803, Inquiry into the consistency of Dr Hutton's Theory of the Earth, Royal Irish Academy, Transactions, 9, 429-487

RICHMOND, G.M. et al., 1965, INQUA Guidebook for Field Conference E, Northern and Middle Rocky Mountains.

RITCHIE-CALDER, Lord, 1982, The Lunar Society of Birmingham, Sci. Amer., 246(6), 136-45

RITTER, D.F., 1978, Process Geomorphology, W.C.Brown:Dubuque

ROBINSON, E. & McKIE, D. (eds), 1970, Partners in Science: Letters of James Watt and Joseph Black, Harvard Univ. Press:Cambridge

ROBISON, J., 1824, Article on 'Rivers' in Encyclopaedia Brittanica, 6th Edition, (see also 3rd ed. 1797)

ROGERS, H.D., 1835, On the Falls of Niagara and the reasoning of some authors respecting them, AJS, 27(2), 326-335

ROGLIC, J., 1972, Historical review of morphologic concepts, in HERAK, M. and SPRINGFIELD, V.T. (eds), 'Karst: important karst regions of the northern hemisphere, Elsevier: Amsterdam, 1-18

RUBEY, W.W., 1933, Equilibrium conditions in debris-laden streams, Amer. Geophys. Union, 14th Annual Mtg, 497-505

---, 1938, The force required to move particles on a stream

bed, USGSPP, 189E, 121-141

RUDBERG, S., 1962, A report on some field observations concerning periglacial geomorphology and mass movement on slopes in Sweden, Biuletyn Periglacjalny, 11, 311-323

RUHE, R.V., 1975, Geomorphology, Houghton Mifflin:Boston

RUSSELL, I.C., 1897, River development as illustrated by the rivers of North America, John Murray:London

RUTIMEYER, L., 1869, Ueber Thal-und Seebildung, Basle

RUXTON, B.P. & BERRY, L., 1959, The basal rock surface on weathered granitic rock, PGA, 70, 285-290

SAID, R., 1950, Geology in tenth century Arabic literature, AJS, 248, 63-66

SALISBURY, R.D., 1924, Physiography (3rd ed.), John Murray:London

SANDERS, E.W., 1921, The cycle of erosion in a karst region (after Cvijic), GR, 11, 593-604

SAUSSURE, H.B. de, 1779, 1786 & 1796, Voyages dans les Alpes (4 vols, last two 1796), Neuchatel

SAVIGEAR, R.A.G., 1952, Some observations on slope development in South Wales, IBGT, 18, 31-51

---, 1956, Techniques and terminology in the investigation of slope form, Premier Rapport Comm. pour l'Etude des versants, IGU, 66-75

---, 1962, Some observations on slope development in north Devon and north Cornwall, IBGT, 31, 23-42

SAWICKI, L., 1909, Ein beitrage zum geographischen Zylus in Karst, Geogr. Zeitsch., 15, 187-204 & 259-81

SCHEIDEGGER, A.E., 1962, Theoretical Geomorphology, Springer Verlag:Berlin (2nd ed. 1970)

SCHEUCHZER, J.J., 1708, Piscium Querelae et Vindiciae, Tiguri.

---, 1723, Itinera per Helvetiae Regiones Alpinas, Leiden

SCHOFIELD, R., 1963, The Lunar Society of Birmingham, Oxford Univ. Press:Oxford

SCHULTZ, S., 1983, The debate over multiple glaciation in the United States: T.C.Chamberlin and G.F.Wright 1889-1899, Earth Science History, 2, 122-129

SCHUMM, S.A., 1963, The disparity between the present rates of erosion and orogeny, USGSPP, 454-H

---, 1973, Geomorphic thresholds and complex response of drainage systems, in MORISAWA, M. (ed.), 1973, Fluvial Geomorphology, George Allen & Unwin:London, 299-310

---, 1976, Episodic erosion: A modification of the geomorphic cycle, in MELHORN, W.N & FLEMAL, R.C. (eds), 1975, 67-85

---, 1977, The Fluvial System, John Wiley:New York

---, 1979, Geomorphic thresholds: the concept and its applications, IBGT, 4ns, 485-515

SCHUMM, S.A. & HADLEY, R.F., 1957, Arroyos and the semi-arid cycle of erosion, AJS, 255, 161-274

SCHUMM, S.A. & LICHTY, R.W., 1965, Space, time and causality
     in geomorphology, AJS, 263, 110-119
SCHYTT, V., 1956, Lateral drainage channels along the
     northern side of the Moltka Glacier, northern Greenland,
     Geogr. Annal., 38, 64-77
SCOTT, W.B., 1897, Introduction to Geology, MacMillan:New
     York
SCROPE, G.P., 1827, Memoir on the Geology of Central France,
     Longman et al.:London
---, 1830, Review of the 1st edition of Lyell's 'Principles
     of Geology,' Quarterly Review, 43, 411-69
SEDGWICK, A., 1831,Presidential Address to the Geological
     Society, 18 Feb. 1831, published in Trans. Geol. Soc,
     Lond. (1826-33), 281-316
---, 1842, Geology of the Lake District in letters addressed
     to W.Wordsworth, Esq., Letter 1
SELBY, M.J., 1970, Slopes and Slope Processes, New Zealand
     Geog. Soc:Waikato
SENECA, L.A., (1871) Natural Questions, Eng. trs. by
     CORCORAN, T.H., Harvard Univ. Press:Cambridge
SHALER, N.S., 1874, Preliminary report on the recent changes
     of level on the coast of Maine, Mem. Bost. Nat. Hist.
     Soc., 2, 321-340
---, 1890, Aspects of the Earth, Smith, Elder & Co.:London
SHARP, C.F.S., 1938, Landslides and related phenomena,
     Columbia University Press:New York
SHEPARD, F.P., 1950a, Longshore bars and longshore troughs,
     BEB Tech. Mem., 15, 1-31
---, 1950b, Beach cycles in Southern California, BEB Tech.
     Mem., 20, 1-20
---, 1963, Submarine Geology, Harper & Row:New York
SHEPERD, R.G. & SCHUMM, S.A., 1974, Experimental study of
     river incision, GSAB, 85, 257-268
SIEGFRIED, R. & DOTT, R.H. (jr) (eds), 1980, Humphry Davy on
     Geology, The 1805 lectures for a general audience, Univ.
     Wisconsin Press:London
SILLIMAN, B., 1839, Appendix; suggestions relative to the
     Philosophy of Geology, in BAKEWELL, R. (sr), 1839
SIMON, L.J., 1957, Additional note on the use of geologists
     in the European Theatre of operation, GSAB, 68,1567
SIMONS, M., 1962, The morphological analysis of landforms: A
     new review of the work of Walther Penck, IBGT, 31, 1-14
SISSONS, J.B., 1958, The deglaciation of part of East
     Lothian, IBGT, 25, 59-77
---, 1962, A re-interpretation of the literature on
     late-glacial shorelines in Scotland with particular
     reference to the Forth area, Trans. Edin. Geol. Soc.,
     19(1), 83-99
---, 1967, Evolution of Scotland's Scenery, Oliver and
     Boyd:Edinburgh
---, 1982, The so-called high interglacial rock shoreline of

Western Scotland, IBGT, 7ns, 205-216

SISSONS, J.B. & CORNISH, R., 1982, Rapid localized glacio-
isostatic uplift at Glen Roy, Scotland, Nature, 297,
213-214

SLAYMAKER, O. & CHORLEY, R.J., 1964, The vigil network
system, J. Hydrol., 2, 19-24

SMITH, D.I. & ATKINSON, T.C., 1976, Process, landform and
climate in limestone regions, in DERBYSHIRE, E. (ed.),
1976, Geomorphology and Climate, John Wiley & Son:London,
367-410

SMITH, H.T.U., 1941, Aerial photographs in geomorphic
studies, J. Geomorphology, 4, 171-205

SMITH, W., 1815, Geological map of England and Wales, with
part of Scotland; exhibiting the Collieries, Mines, and
Canals, the Marshes and Fenlands originally overflowed by
the Sea; and the Varieties of Soil, according to the
variations of the Substrata, by William Smith, Mineral
Surveyor, London

SNYDER, C.T., 1957, Use of geologists in planning the
Normandy invasion, GSAB, 68, 1565

SORBY, H.C., 1850, On the excavation of the valleys in the
Tabular Hills, as shown by the configuration of
Yedmandale, near Scarbro', Proc. York. Geol. Soc., 3,
169-72

SPARKS, B.W., 1960, Geomorphology, Longmans:London

---, 1971, Rocks and Relief, Longmans:London

SPREITZER, H., 1932, Zum Problem der Piedmonttreppe, Mitt.
Geogr. Ges. Wien, 75, 327-64

SPRUNT, B., 1973, Digital simulation of drainage basin
development, in CHORLEY, R.J. (ed.), 1973, Spatial
Analysis in Geomorphology, Methuen:London, 372-375

STANLEY, H.M., 1876, Letters of Mr H.M.Stanley on his Journey
to Victoria Nyanza and Circumnavigation of the Lake,
Proc. Royal Geog. Soc., XX, 134-159

STARKEL, L., 1976, The role of extreme (catastrophic)
meteorological events in the contemporary evolution of
slopes, in DERBYSHIRE, E., (ed.) 1976, Geomorphology and
Climate, John Wiley & Sons:London, 203-246

STEERS, J.A., 1926, Orford Ness, a study in coastal
physiography, PGA, 37, 306-325

---, 1927, The East Anglian Coast, GJ, 69, 24-48

---, 1945, Coral reefs and air photographs, GJ, 106, 233-235

---, 1946, The Coastline of England and Wales, Cambridge
Univ. Press:Cambridge (revised and expanded 2nd ed. 1964)

---, 1953a, The Sea Coast, Collins:London

---, 1953b, The East Coast Floods, GJ, 119, 280-98

STEERS, J.A. & SMITH, D.B., 1956, Detection of movement of
pebbles on the sea floor by radioactive methods, GJ, 122,
343-345

STEINBERG, S.H., 1966, Five hundred years of printing,
Penguin:Harmondsworth

STENO, N., 1669, De solido intra solidum naturaliter contento
    dissertationis prodomus, Florentiae, (Eng. trs.,
    OLDENBURG, H., 1669, also Eng. trs. by WINTERS, J.G.,
    1916, MacMillan:New York, rprt, WHITE, G.W. (ed.), 1968,
    Hafner:New York)
STODDART, D.R., 1960, Colonel George Greenwood, the father of
    modern subaerialism, SGM, 74, 108-110
---, 1965, in discussion at the Royal Geographical Society of
    'Denudation in Limestone regions, a symposium,' GJ, 131,
    34-57
---, 1969, Climatic geomorphology: a review and
    re-assessment, Prog. in Geog., 1, 159-222
---, 1975, 'That Victorian Science:' Huxley's Physiography
    and its impact on geography, IBGT, 66, 17-40
STORRIE, M., 1969, William Bald, F.R.S.E. c1789-1857,
    Surveyor, Cartographer and Civil Engineer, IBGT, 47,
    205-231
STRABO, (1969), The Geography of Strabo, Eng. trs. JONES,
    H.L. & STERRETT, R.S., Harvard Univ. Press:Cambridge
STRAHLER, A.N., 1950a, Equilibrium theory of erosional slopes
    approached by frequency distribution analysis, AJS, 248,
    673-96 & 800-14
---, 1950b, Davis' concept of slope development viewed in the
    light of recent quantitative investigations, AAAG, 40,
    209-13
---, 1952, The dynamic basis of geomorphology, GSAB, 63,
    923-38
SUESS, E., 1883-1908, Der Antlitz der Erde, F. Tempsky:Wien,
    3 volumes (French trs. by DE MARGERIE, E., 1897-1918, La
    face de la terre, Armand Colin:Paris, English trs. by
    SOLLAS, H.B.C. & SOLLAS, W.J., 1908-1924, The face of the
    earth, Clarendon Press:Oxford)
SUGDEN, D.E., 1977, Reconstruction of the morphology,
    dynamics and thermal characteristics of the Laurentide
    Ice Sheet at its maximum, Arctic & Alp. Res., 9, 21-47
---, 1978, Glacial erosion by the Laurentide Ice Sheet, J.
    Glaciol., 20, 367-91
SUGDEN, D.E. & JOHN, B.S., 1976, Glaciers and Landscape,
    Arnold:London
SUNDBORG, A., 1956, The river Klaralven: a study in fluvial
    processes, Geogr. Annaler, 38, 127-316
SUPAN, A., 1884, Grundzuge der physischen Erdkunde, Leipzig
SWEETING, M.M., 1943, Wave trough experiments in beach
    profiles, GJ, 101, 162-72
---, 1950, Erosion cycles and limestone caverns in the
    Ingleborough district, GJ, 115, 63-78
---, 1964, Some factors in the absolute denudation of
    limestone terrains, Erdkunde, 18, 92-95
---, 1972, Karst Geomorphology, Columbia Univ. Press:New York
SWIFT, D.J.P., 1975, Barrier-Island genesis: evidence from
    the central Atlantic shelf, eastern USA, Sed. Geol., 14,

1-43
SWINNERTON, A.C., 1932, Origin of limestone caverns, GSAB, 43, 663-694

TARGIONI-TOZZETTI, G., 1752, Realizion d'alcun viaggi fatti in diverse parti della Tuscana:Florence

TARR, R.S., 1898, The peneplain, Amer. Geol., 21, 351-70

TARR, R.S. & MARTIN, L., 1914, Alaska glacier studies, NGS:Washington

TAYLOR, R., 1794, An account of repeated shocks of earthquakes felt in Perthshire, in a letter to the Reverend Mr Finlayson, F.R.S.Edin. from Mr RALPH TAYLOR, (dated Jan. 19 1790) and postscript added Jan. 24 1793, RSET, 3, 240-246.

THOMAS, M.F., 1978, Denudation in the tropics and the interpretation of the tropical legacy in higher latitudes - a view of the British experience, in EMBLETON, C. et al. (eds), 1978, 185-202

THOMSON, J., 1817, A New General Atlas, Thomson:Edinburgh

THORBECKE, F., 1927a, Die Formenschatz im periodisch Trockenen Tropenklima mit uberwiegender Regenzeit, Duss. Geogr. Vortage und Erorterungen 3, 10-17 (Eng. trs. in DERBYSHIRE, E. (ed.), 1973, 96-103)

---, 1927b, Morphologie de Klimazonen, Dusseldorf Geogr. Vortrage u. Erorterungen

THORN, C.E., 1979, Bedrock freeze-thaw weathering regime in an Alpine environment, Colorado Front Range, ESP, 4, 211-228

---, (ed.), 1982, Space and Time in Geomorphology, George Allen & Unwin:London

THORNBURY, W.D., 1954, Principles of Geomorphology, John Wiley:New York (2nd ed. 1969)

---, 1965, Regional Geomorphology of the United States, John Wiley:New York

THORNES, J.B., 1977, Hydraulic geometry and channel changes, in GREGORY, K.J. (ed.), 1977a, 91-100

---, 1978, The character and problems of theory in contemporary geomorphology, in EMBLETON, C. et al. (eds), 1978, 14-24

---, 1980, Structural instability and ephemeral channel behaviour, ZfG, 36nf (supp.), 233-244

THORNES, J.B. & BRUNSDEN, D., 1977, Geomorphology and Time, Methuen:London

TINKLER, K.J., 1983a, On Hutton's authorship of an 'Abstract of a dissertation..concerning the System of the Earth, its Duration and Stability,' GM, 120, 631-4

---, 1983b, John Playfair on a glacial erratic in the Isle of Arran, Scot. J. Geol., 19, 129-134

TRICART, J., 1968, Geomorphologie Structurale, Paris (Eng. trs. BEAVER, S.H. & DERBYSHIRE, E, 1974, Longman:London)

TRICART, J. & CAILLEUX, A., 1965, Introduction a la geomorphologique climatique, S.E.D.E.S.:Paris (Eng. trs. De JONGE, C.J.K., 1972, Longman:London)

TRIMMER, J., 1832, On the diluvial deposits of Caernarvonshire, Proc. Geol. Soc., 1(22), 331-2

TROLL, C., 1948, Der subnivale oder periglaziale Zyklus der Denudation, Erdkunde, 2, 1-21

TUAN, Y.F., 1958, The misleading antithesis of Penckian and Davisian concepts of slope retreat in waning development, Proc. Indiana Acad. Sci., 67, 212-14

TYLOR, A., 1869, On the formation of deltas; and on the evidence and cause of great changes in the sea level during the Glacial Period, QJGS, 22, 463-8

---, 1875, Action of denuding agencies, GM, 22, 433-73

TYNDALL, J., 1860, The Glaciers of the Alps, London

UNITED STATES GEOLOGICAL SURVEY, 1974, A brief History of the U.S. Geological Survey, USGS:Washington

VANDERPOOL, N.L., 1982, ERODE- A computer model of drainage basin development under changing baselevel conditions, in CRAIG, R. & CRAFT, R.L. (eds), 1982, 214-223

VARENIUS, B., 1672, Geographia Generalis, (NEWTON, I., ed.), Cambridge

VENETZ, I., 1821, verbal presentation, see VENETZ, I., 1833

---, 1829, in Actes Soc. Helv. Sc. Nat. 15eme session, page 31

---, 1833, Memoir sur les variations de la temperature dans les Alpes de la Suisse, Denlschriften derallg. schweiz. Geseel. ges. Naturwissenschaften, 1, 1-38 (includes material from 1821)

VITA-FINZI, C.,1969, The Mediterranean Valleys, Cambridge Univ. Press:Cambridge

VON ENGELN, O.D., 1942, Geomorphology, systematic and regional, MacMillan:New York

WALKER, H.J., 1978, Research in coastal geomorphology: basic and applied, in EMBLETON, C. et al. (eds), 1978, 203-223

WALLING, D.E., 1977, Suspended sediment and solute response characteristics of the River Exe, Devon, England, in DAVIDSON-ARNETT, R. & NICKLING, W. (eds), 1978, Research in Fluvial Geomorphology, Geo Books; Norwich, 169-197

WARREN, E., 1690, Geologia, or a discourse concerning the earth before the deluge, London

WASHBURN, A.L., 1956, Classification of patterned ground and a review of suggested origins, GSAB, 67, 823-825

WATERS, R.S., 1958, Morphological Mapping, Geography, 43, 10-17

WAYLAND, E.J., 1920, Annual Report of the Geological Survey
    of Uganda, for the year ended 31 March 1920
---, 1933, Peneplains and some other erosional platforms,
    Ann. Rept Bull. Protectorate of Uganda, Geol. Surv. &
    Dept. of Mines, Note 1, 77-79
---, 1934, Rifts, Rivers, Rain and Early Man in Uganda, J.
    Roy. Anthrop. Inst., 64, 333-352
WEBB-SEYMOUR, Lord, 1815, An account of observations, made by
    Lord Webb-Seymour and Professor Playfair, upon some
    geological appearances in Glen Tilt and adjacent country,
    drawn up by Lord Webb-Seymour, RSET, 7, 303-376
WEGENER, A., 1915, Die Enstehung der Kontinente und Ozeane,
    (later editions 1920, 1922, 1929) F. Vieweg &
    Sohn:Braunschweig , (Eng. trs. of 1929 edition by Biram,
    J., 1966, The Origin of the Continents and the Oceans,
    Dover:New York)
WHEWELL, W., 1831, Review of 'Principles of Geology', British
    Critic, 9, 180-206
WHISTON, W., 1696, A new Theory of the Earth, London (rprt,
    1978, Arno Press:New York)
WHITAKER, W., 1867, On subaerial Denudation, and on Cliffs
    and Escarpments of the Chalk and Lower Tertiary Beds, GM,
    4, 447-454 and 483-493
WHITE, G.W., 1951a, Lewis Evans' contribution to early
    American Geology 1743-1755, Ill. Acad. Sci. Trans., 44,
    152-158 (rprt, in WHITE, G.W., 1978)
---, 1951b, Lewis Evans' early American notice of isostasy,
    Science, 114, 302-3 (rprt, WHITE, G.W., 1978)
---, 1978, Essays on the History of Geology, foreword by
    ALBRITTON, C.C., Arno Press:New York
WHITE, S.E., 1976, Is frost shattering really only hydration
    shattering? A review, Arctic and Alpine Research, 8, 1-6
WHITEHURST, J., 1778, An inquiry into the original state and
    formation of the Earth, W. Bent:London (rprt, of 2nd ed.
    1786, 1978, Arno Press:New York)
WILLIAMS, J., 1789, The Natural History of the Mineral
    Kingdom, Edinburgh
WILLIAMS, J.D. & MEADE, R.H., 1983, World-wide delivery of
    river sediment to the oceans, JG, 91, 1-22
WILLIAMS, P.J., 1957, Some investigations into solifluction
    features in Norway, GJ, 123, 42-58
---, 1959, An investigation into processes occurring in
    solifluction, AJS, 257, 481-490
---, 1966, Downslope soil movement at a sub-arctic location
    with regard to variations of depth, Canadian Geotechnical
    Journal, 3, 191-203
WILLIAMS, P.W., 1966, Morphometric analysis of temperate
    karst landforms, Irish Speleology, 1, 23-31
---, 1972, Morphometric analysis of polygonal karst in New
    Guinea, BGSA, 83, 761-796
WILLIAMS, R.B.G., 1964, Fossil patterned ground in Eastern

England, Biuletyn Periglacjalny, 14, 337-349

WILLIAMS, R.B.G. & ROBINSON, D.A., 1981, Weathering of sandstone by the combined action of frost and salt, ESPL, 6, 1-9

WILLIAMS, W.W., 1947, The determination of gradients on enemy-held beaches, GJ, 9, 76-93

---, 1953, La tempete des 31 Jan. et 1 Feb. 1953, Bull. Inform. C.O.E.C., 5, 206-10

---, 1956, An East Coast Survey (discussion), GJ, 122, 317-34

---, 1960, Coastal Changes, Routledge & Kegan Paul:London

WILLIAMS, W.W. & KING, C.A.M., 1951, Observations faites sur la plage des Larentes en Aout, 1950, Bull. Inform. C.O.E.C., 2, 363-8

WILSON, L.G., 1972, Charles Lyell the years to 1841, The revolution in Geology, Yale Univ. Press:New Haven & London

WOLMAN, M.G. & LEOPOLD, L.B., 1957, River flood plains; some observations on their formation, USGSPP, 282-C

WOLMAN, M.G. & MILLER, J.P., 1960, Magnitude and frequency of forces in geomorphic processes, JG, 68, 54-74

WONG, S.T., 1963, A multivariate statistical model for predicting mean annual flood in New England, AAAG, 53, 298-311

WOODWARD, J., 1695, Essay towards a Natural History of the Earth, Wilkin:London

WOOLDRIDGE, S.W., 1958, The trend of geomorphology, IBGT, 25, 29-35

WOOLDRIDGE, S.W. & LINTON, D.L., 1939, Structure, surface and drainage in south-east England, IBGT, 10, (revised rprt 1955, rprt 1964)

WOOLDRIDGE, S.W. & MORGAN, R.S., 1937, The physical basis of geography: An outline of geomorphology, Longmans:London (2nd ed., 1959, with title reversed)

WOOLNOUGH, W.G., 1930, The influence of climate and topography in the formation and distribution of products of weathering, GM, 67, 123-132

WRIGHT, G.F., 1889, The Ice Age in North America, D. Appleton & Co.:New York

WRIGHT, W.B., 1914, The Quaternary Ice Age, MacMillan:London (2nd ed. 1937)

YATES, J., 1830-1, On the formation of alluvial deposits, Proc. Geol. Soc., 1, 237-9

YATSU, E., 1966, Rock control in geomorphology, Sozosha:Tokyo

YOCHELSON, E.L. (ed.), 1980, The scientific ideas of G.K.Gilbert, GSA Spec. Pap., 183

YOUNG, A., 1960, Soil movement by denudation processes on slopes, Nature, 188, 120-22

---, 1961, Characteristic and limiting slope angles, ZfG, 5nf, 126-131

---, 1963, Deductive models of slope evolution, Nach. Akad. Wiss. Gottingen ii. Math-Phys., Klasse 5, 53-66

---, 1971, Slope profile analysis: the system of best units, IBG Spec. Pap., 3, 1-14

---, 1972, Slopes, Oliver and Boyd:Edinburgh

---, 1978, Slopes 1970-1975, in EMBLETON, C. et al. (eds), 1978, 73-83

ZITTEL, K.A. von, 1901, History of Geology and Palaeontology (Eng. trs. by Marie M. OGILVIE-GORDON), Walter Scott:London (rprt, 1962, J.Cramer:Weinheim, Wheldon & Wesley and Hafner:London & New York)

ZOTIKOV, I.A., 1963, Bottom melting of the central zone of the ice shield of the Antarctic continent, Bull. Int. Ass. Sci. Hydrol., 8, 36-43

INDEX

Mill, H.R. 151
Miller, H. 62
Miller, J.P. 182, 188, 200,
 203, 233
Milne, A.A. 185
mining 67, 68
 hydraulic 143
Town and Country Planning
 Ministry 179
Miocene 214
misfit
 non-climatic 224
Mississippi 85, 87, 88, 133,
 135, 140
 age of delta 137
 Humphreys & Abbott 138
Missoula, Lake 169
Mitchill, S.L. 136
modnadnocks 148
Moel Tryfan 113
Montlosier, F. de 36, 40
Montmorenci Falls 98
Moon 14, 177
Morgan, R.S. 151, 162, 234
Moro, L. 32, 33
morphoclimatic regions 156,
 161
morphogenetic regions 156
Morph. Analysis of Landforms
 difficult German 166
 English trs. 166
 mis-interpreted 166
morphological mapping 175,
 220
morphologists 174
morphometry
 Baulig 183
 Clarke 183
 climatic geomorphology
  ignored 184
 De Martonne 183
 De Smet 183
 Flint, J-J. 215
 French 183
 German 183
 Gregory, K.J. 215
 karst
  Stoddart 196
  Williams, P.W. 196
 network 184

Patton, Baker 215
Peguy 183
 scale 183
Mortensen, H. 160
Mosaic Deluge (see Flood)
Mt. Blanc 56, 60, 120
 Davy 62
 Jura erratics 121
 Saussure ascends 79
Muir, J. 191
Muller, S.W. 194
Murchison, R.I. 74, 114, 123,
 125, 128
 Auvergne valleys 85
 Bayfield's letter 113
 catastrophist 112
 craters of elevation 113
 defines drift 107
 denudations 112
 depositional environments
  112
 distinguishes drifts 112
 drift for diluvium 112
 Geological Survey 116
 glaciation 'peculiar' 113
 influence on Hall 92
 local drift 113
 neo-diluvialist 112, 116
 opposes Buckland 111
 refutes Flood 86, 112
 Russian Drift 113
 weak modern rivers 113
 Woolhope Dome 113
Murray, J. 115, 136
 accepts denudation 62
 denies restoration 62
Mycenae 21

N

Nace, R.L. 174
Nansen, F. 134
Nanson, G.C. 6
National Air Photo. Library
 Ottawa 175
Natural Env. Res. Council 204
Nature Conservancy 180
Naumann, C.F. 4
neocatastrophism
 Baker & Patton 212

304